VEGETABLE CROPS BREEDING

VEGETABLE CROPS BREEDING

Ravindra Mulge
Dean
College of Horticulture
(University of Horticultural Sciences, Bagalkot, Karnataka, India)
Halladakeri Farm, Hyderabad Road
Bidar 585 403, Karnataka, India

CRC Press is an imprint of the
Taylor & Francis Group, an **informa** business

First published 2021
by CRC Press
2 Park Square, Milton Park, Abingdon, Oxon, OX14 4RN

and by CRC Press
6000 Broken Sound Parkway NW, Suite 300, Boca Raton, FL 33487-2742

CRC Press is an imprint of Informa UK Limited

British Library Cataloguing-in-Publication Data
A catalogue record for this book is available from the British Library

Library of Congress Cataloging-in-Publication Data
A catalog record has been requested

ISBN: 978-0-367-75480-8 (hbk)

Dr. K. M. INDIRESH
Vice-Chancellor
UHS, Udyanagiri
Navanagar, Bagalkot-587 104

Foreword

Our country is endowed with a diverse climate ranging from tropical to temperate that is possible to grow an array of horticultural crops in one or other parts of India. Among the horticulture crops, vegetables are easily accessible and cheaper sources of vitamins, minerals and other elements essential for human being. The role of vegetables in combating malnutrition and deficiency disorders in human being is well conceived now as days. Looking into the importance of vegetables, the breeding and genetics in general has remained neglected particularly in our country. Even though we are second largest producers in the world next to China, the domestic demand for vegetables will rise from 220 million tons to 375 million tons by 2033. There is a great need to bring more area under vegetable seed crops by adopting modern breeding technology especially temperate, tropical and sub-tropical types which have direct bearing for the improvement in socio economic status of farming community.

The book **Vegetable Crops Breeding** may be of great utility and intend to be a better guide to students, teachers, research scientists, various seed companies, vegetable seed producers and farming community as whole who directly and indirectly are engaged in the seed trade and are responsible for the development of vegetable seed industry.

I appreciate and congratulate Dr. Ravindra Mulge, Dean, College of Horticulture, Bidar in writing this book of immense value for all interested in breeding of vegetable crops. The theoretical and practical information contained in the form of different chapters on various aspects will be of great use to all interested.

Place : UHS, Bagalkot
Date : 30.11.2019

(K.M. Indiresh)
Vice Chancellor

Preface

Vegetables are important constituents of our healthy diet and provide more profits to the growers. Crop improvement is the most satisfying method of increasing the productivity of any crop. With the experience of teaching vegetable breeding to graduate and post graduate students for more than fifteen years and involvement in developing few varieties of vegetable crops and serving as Professor and Head of the Department of Vegetable Science, idea of bringing out a book was conceived and started compiling notes during all these years. Listing of references is not exhaustive to avoid bulkiness of the book where information is presented in very comprehensive manner.

Production and consumption of vegetables has expanded greatly. Improved varieties have had a main role in the increase in yield and quality of vegetable crops. The diversity of vegetable crops is appalling and greatly contributed to increased production and consumption of vegetables world over. For improvement of crops information on origin, distribution and evolution of crop and its related species is very essential. Information on genetics and genetic resources is prerequisite to choose the appropriate breeding strategies to fulfill the objectives. Objectives of breeding vary with region and also purpose for which the product is used i.e., for fresh market, for processing or dual purpose and how the crop is grown i.e., under protection, open field cultivation or kitchen garden etc. Depending on the objectives and genetic resources, breeding methods or procedures can be adopted and it will results in useful varieties or hybrids. Information on breeding of vegetables crops is covered in very abridged form. More number of crops covered in this book by appending and updating information on genetics, genetic resources, breeding methods/procedures and varieties developed in each of the 26 vegetable crops wherever information is available. Origin and evolution, genetic resources, genetics of fruits, breeding methods and varieties/hybrids developed is the sequence followed in presenting the information on each of the crops. It can be the reference book for teachers and graduate and post graduate students in Vegetable Science or Horticulture or Agriculture. I am thankful to Dr. Parvati Pujar and Mrs. Neelambika and others who helped in editing this manuscript.

Author

Contents

Introduction to Vegetable Crop Breeding

Vegetables are important constituents of nutritional security and economic prosperity due to their short duration, high productivity, nutritional richness, high profitability and high employability. Breeding is probably as old as cultivation of crops itself. Earlier simple selection was followed and later new breeding methods were developed, including hybridization techniques, culminating with the use of recently developed molecular tools, all leading to our modern improved vegetable cultivars. There are more than hundred vegetable crops cultivated worldwide but only few of them have attracted commercial breeding attention. Improved hybrids or open pollinated vegetable cultivars will continue to be the most effective, environmentally safe and sustainable way to ensure global food security in the future. Information on floral biology and selfing and crossing techniques are very basic requirements for breeding of vegetable crops.

Floral Biology of Vegetable Crops

Information on floral biology of the crop is very essential to know the gene flow and also for breeding manipulations through selfing and crossing. Floral biology includes study of the flowers and their developments which include opening of flower (anthesis), dehiscence of anthers, pollen viability, stigma receptivity, etc. Floral biology is influenced by environmental factors. Usually anthesis of flower takes place in the morning hours in most of the crops except few crops like bottle gourd, ridge gourd, snake gourd where anthesis takes place afternoon or evening. Information on the floral biology of vegetable crops is summarised below. Timings of 24 hours pattern is considered and 0 indicates on the day of anthesis, bA indicates day/hour(hr) before anthesis and AA indicates day/hour after anthesis.

Crop	Anthesis	dehiscence	Pollen fertility	Stigma receptivity	Place of report
Tomato	7 to 8	9 to 11	0	16hr 1D	India, Hissar
Potato	5 to 6	5.30 to 7	0+next day morning	0 + next day afternoon	USA
C. annuum	5 to 6	8 to 11	0	0	India
Brinjal	6 to 11.307 to 14	4.15 to 11 hr 7.15 to 14 hr	0+2 days AA0	-2 to 3 hr bA to 2 days AA -1 to 2 days AA	New Delhi, Bihar, India
Okra	8 to 10.30	6 to 11	6 to 12 hr	-1 to 0	Jaipur, India
Peas	5.45 to 6.30	5.45 to 7.30	0 to 24hr	-48 to 24 hr	Bangalore
Dolichos bean	9 to 17 hr	5 to 14 hr	0	0	Nagpur India
Sweet potato	5 to 7	5 to 7.30	0	8 to 10hr fall 6 to 9hr spring	USA
Onion	6 to 7 hr	7 to 17 hr next d also	0	0	USA
Radish	8 to 10 hr	8 to 10.30 hr	0	2 to 3 days BA to day after A	Japan
Cabbage	8 to 10 hr	8 to 10.30 hr	0	2 to 3 days BA to day after A	Japan
Musk melon	5.30 to 6.30 hr	5 to 6 hr	5 to 14 hr	2 hr BA to 2 -3 hr AA	India
Water melon	6 to 7.30 hr	5 to 6.30 hr	5 to 11 hr	2 hr BA to 2 -3 hr AA	India
Bottle gourd	17 to 20 hr	13 to 14.30 hr	0	36 hr BA to 60 hr AA	India
Snake gourd	18 to 21 hr	Shortly before A	10 hr BA to 49hr AA	7 hr BA to 51 hrs AA	-
Ridge gourd	17 to 20 hr	17 to 20 hr	0 to 2-3 d AA	6 hr BA to 84 hr AA	-
Cucumber	5.30 to 7.00	4.30 5 hr	Up to 14 hr	12 hr BA to 6 -7 hr AA	India
Cucurbita pepo	3.30 to 6.00 hr 2.20 to 3.45 hr	21 hr to 3 hr -	16 hr AA -	2 hr BA to 10 hr AA Through out opening period	Ludhiana India Mysore, India

Natural cross pollination in vegetable crops

Crop	Feature	NCP(%)	Classification	Pollination vector	Report place
Tomato	-	0.59-4.9	Self	Bumble bee	USA
Brinjal	Heterostyly	0.2-46.8	Often cross	Insect & wind	Japan
Capsicum annuum	Protandry	7-37	Cross/often cross	Honey bees & thrips	USA
,,		58-68	,,	Insect	India
Potato		0-20	Often cross	Bumble bees	Scotland
Cow pea		1	Self	Bumble bees	USA
Cluster bean		1-4.4	Self	Bees	USA
Broad bean		0-46	Self/cross		India
French bean		6-10	Self		Brazil
Lima bean		0-0.53	Self		
Soybean		0.07-0.18	Self		USA
Celery		47-87	Cross		USA
Okra	protogyny	0.34-27.3	Often cross		
Sweet potato	Self incomp.	2.21-56.41	Cross		
Spinach	Separation of sex	19.6-96.8	Cross	Wind	
Carrot	Protandry	97.6-98.9	Cross		USA
Radish	S.I.	highly	Cross	Bumble bees	
Cauliflower	S.I.	40-50	Cross	Honey, bumble bees & blow flies	
Cabbage	S.I.	73	Cross	,,	
Onion	Protandry & MS	95-100	Cross	Blowflies	
Musk melon	Protandry, monoecious,	5.4-98	Cross	Honey bees	USA
Cucumber	Separation of sex	65-70	Cross	Honey bees	

Isolation distance: (based on extent of out crossing)

0-10%:100-150m, 11-25%:200-300m,
25-50%:300-500m 50-75%:500-1000 m,
75-100%:1000 m

Isolation distance: (based on extent of out crossing)

0-10%: 100-150 m 11-25%: 200-300 m

25-50%: 300-500 m 50-75%: 500-1000 m

75-100%: 1000 m

Selfing and Crossing Techniques

The artificial hybridisation process consists of following steps:

1. Study of floral morphology and emasculation
2. Bagging and protection of flowers of the selected male and female parents.
3. Pollination of the emasculated flower.
4. Records and labelling.
5. Harvesting and observation

The instruments commonly used in emasculation and crossing are needle, scalpel, forcep, scissor, bag, pocket lens, tag and label, U-pin, pencil, spirit (alcohol) and brush.

Tomato

For selfing an unopened flower bud should be covered in a small butter paper bag and threads can be tied for identification of selfed fruit.

For crossing, emasculation is usually done one day prior to anthesis. At this stage, the sepals have started to separate and anthers and corolla are beginning to change from light to dark yellow. The stigma is fully receptive at this stage allowing for pollination even immediately after emasculation. Anthers are removed as a group with or without the surrounding corolla by inserting forceps between the sepals to grip the base of the anthers and/or petals which are then removed by a firm but steady pull without damaging the style and stigma. Pollen can be collected from a mature flower of known origin by tearing the anther cone and gently tapping it on to the thumb nail. Stigma of emasculated flower is gently brushed on to the thumb nail where pollen is collected to facilitate smearing of pollen on stigma. Protection of pollinated flowers by covering it with small butter paper bag is essential. Pollinated flower should be tagged with the label carrying information about male parent and date. With change in parent or plant sterilise forceps and thumb nail by dipping them in alcohol or rectified spirit.

Brinjal

For selfing, select a healthy long or medium styled, well-developed bud from the central portion of the plant and pollen extracted from same plant can be smeared on the stigma with the help of forceps after forcefully opening the bud. Cover the pollinated flower with butter paper bags to avoid mixtures and label each flower by tying the thread.

For crossing, the flower has to be emasculated. A healthy long or medium styled, well-developed bud is selected. The bud is opened gently with the help of fine pointed forceps one day prior to anthesis and all the five anthers are carefully removed without damaging the style and stigma. For pollination, freshly dehiscing anthers are picked and are pierced in to the sides and sufficient pollen can be scooped out with the help of forceps/needle. Pollen powder is smeared on to the stigma of emasculated flower bud. It is labelled and covered with small pollination bag. Sterilisation of needles/forceps is very essential to avoid mixtures.

Hot and Sweet Pepper

For selfing an individual flower can be wrapped loosely in a cotton strip and flower stalk can be tied with thread as a mark of selfing. Entire plant can be covered in cheese cloth supported by an iron frame after removing all the set fruits and opened flowers. After an interval of 20 to 30 days remove the cloth and look for set fruits and label them individually by tying a thread as a mark of selfing.

Crossing can be done any time during the day but morning hours are preferred. Flowers are emasculated in bud stage by removing the anthers individually after opening the fully developed flower. Ensure, non-bursting of anthers in the unopened flower bud. Pollen is transferred to the stigma either from mature undehisced anthers by scooping it out through the lateral sutures with the needle or by touching a freshly dehisced anther to the stigma with the forceps. Cover the pollinated flowers with cotton wrapping and label them by tagging a jewellery tags. Sterilisation of forceps and needles before emasculation and pollination is very essential.

French bean

It is a highly self-pollinated crop and for selfing, a flower bud can be covered with paper bag and labelled.

For crossing, emasculation has to be done by selecting an unopened flower bud. The standard is detached from below with a pair of forceps and bent backward. The keel is pulled in pieces with the forceps. Care should be taken to turn in the same direction as the spiral winding, otherwise the style may break. After pulling of keel, the stamens are removed. For the supply of pollen, freshly opened flowers are collected. The thickly pollinated stigmas emerge as soon as the wings are pressed downward. This stamen is rubbed against the stigma of the emasculated bud. Sometimes, the thickly pollinated stigma of the open flower to be used as male is hooked into the stigma of the flower bud to be pollinated. Pollination without emasculation has also been suggested. In this method, the standard is detached and unfolded. By pressing the left-hand side wing downward, the unpollinated stigma is exposed. Pollen is rubbed into this stigma. The emasculation and pollinations are done simultaneously in the forenoons. If part of the style or stigma is injured, the flower drops after 2 or 3 days. Failure in crossing may result in flower drop or pods without seeds or pods of 2 to 3 cm long are formed but they dry-up afterwards.

Cowpea

Bagging of individual flowers with labelling will ensure selfing.

For crossing, emasculation needs to be carried out in mature flower buds in the preceding evening. The flower buds likely to bloom next day (large size, yellowish back of the standard petal) is selected for emasculation. The bud is held between the thumb and the forefinger with the keel side uppermost. A needle is run along with ridge where the two edges of the standard unite. One side of the standard is brought down and secured in position with thumb. Same is done with one of the wings to expose the keel. Exposed keel is slit about 1/16 inch from the stigma. A section of keel is also brought down and secured in position under the thumb. The stamens which are exposed can be removed with pointed forceps. Afterwards, the disturbed parts of standard, wing and keel are brought in original position as far as possible. To prevent drying out of the emasculated bud, a leaflet may be folded and pinned around the bud. A tissue paper can be used to cover and protect the bud. Pollination is done next morning by using the pollen of freshly opened flower. The standard and wings of male flower are removed. By slight depression of the keel, stigma covered with pollen grains protrudes out. This itself can be used as a brush for pollination. Cowpea flowers are highly sensitive and drop off easily with slight mechanical disturbance or injury.

Cucumber

Phases of sex expression and genetics of sex expression and information on floral biology in cucumber will serve as guideline for carrying out selfing and crossing techniques.

Selfing: Induction of maleness in gynoecious lines and femaleness in androecious lines are very important to carry out selfing. Anthers of the male flowers or hermophrodite flowers are brushed against the receptive stigma of the same plant. The staminate flowers are in cluster with short and slender pedicels. The pistillate flowers are usually solitary with stout and short pedicels. Pollinated flower has to be covered with butter paper bags.

Crossing: For crossing purpose, pistillate flowers are closed with a rubber band/ wrapped with cotton pad to protect against unwanted pollen about 1 to 2 days prior to opening. Anthesis takes place around 5.30 to 7.00 hours. Dehiscence occurs around 4.30 to 5.00 hours. Pollen fertility is upto 14 hour. Similarly male flowers are also protected. After 1 to 2 days, pollination is carried out from pollen of protected staminate flower preferably in the morning/forenoon. After about 4 weeks, the fruits are harvested and allowed to further ripening by about another week. For seed collection, the fruits are cut-open and seeds are collected in a glass jar with water and left as such for two days to remove gelatinous mass. The success of controlled pollination may be enhanced by removal of any previously set fruit as first fertilized flower inhibits the development of subsequent fruits. Therefore, controlled pollination should be done as early as possible after flowering begins.

Squash and Pumpkin

The pistillate and staminate flower buds are located a day before anthesis and are protected by tying the tip of corolla tube. The following morning, as soon as the pollen sacs dehisce, the staminate flower is taken out and pollen is applied on the receptive stigma of the pistillate flower. This is easily done by rubbing the anthers against the stigma. Pollination can be done in the morning and can be continued till noon. For better fruitset, it is desirable to pollinate first few pistillate flowers and to remove the previously set fruits by open-pollination. Crossed or selfed fruits should be properly tagged and a stake may be placed adjacent to the pollinated fruit to mark in location.

Radish

Radish is a cross-pollinated crop due to sporophytic system of self-incompatibility. It shows considerable amount of inbreeding depression on selfing. Selfing can be accomplished by bud-pollination. The flower buds are pollinated two days prior to opening by their own pollen by applying fresh pollen from previously bagged flowers of the same plant. Emasculation is not necessary in bud-pollination. After pollination, the buds are to be protected from foreign pollen by enclosing the particular branch bearing these buds in a muslin cloth bag.

In crossing, the same technique is used as in bud-pollination except that in the crossing, the buds of the female parent are emasculated a day prior to opening and are pollinated by pollen collected from the flowers of the male parent which were also bagged before opening. The artificial pollination is done by hand by shaking the pollen over the stigma directly from the freshly opened but previously bagged buds of male parent. However, for large scale F_1 seed production, perfect self-incompability and insect proof cages can be used. Cross-compatible and self-incompatible parents are put in cages with honey bee hives to produce crossed seeds.

Onion

Selfing in onion is done only on a limited scale as it becomes difficult to maintain the inbred lines beyond S_2 generation due to drastic inbreeding depression. Selfing is done by putting individual cages over the plants. Flies are generally used to ensure pollination within cages. Sometimes it is convenient to enclose 2 to 3 umbels of the same plant in a muslin cloth bag before anthesis. After anthesis, the umbels are rubbed against each other daily for a few days to ensure self-pollination.

Crossing: As soon as few buds in an umbel open, the whole umbel of the female parent is bagged in a muslin cloth bag. The flowers are removed daily for a few days until the peak flowering has reached after which buds are emasculated as they open and when sufficient buds have been emasculated, the remaining young flower buds are removed. The umbel of pollen parent covered by a muslin cloth bag is cut off and its stalk placed in a glass bottle filled with water. This bottle is fastened to a bamboo/wooden stake and fixed in soil close to the female parent. Female parent umbel (emasculated one) and the pollen parent umbels are now enclosed in the same common bag. For a few days in the morning, the male umbel is gently rubbed over the emasculated umbel to ensure pollen shedding and cross-pollination. A few common houseflies can also be introduced into the bag for pollen transfer. However, cross-pollination through hand emasculation is extremely difficult and it can be substituted with marker genes.

High level of inbreeding depression is observed in Brussels sprouts, kale, cabbage and broccoli. Moderately to low inbreeding depression is observed in radish, carrot, onion and beet.

1

Tomato

Tomato is relatively recent addition to the world's vegetable crops. Now became one of the most important vegetable crops and is popular & widely consumed. It is being grown extensively in USA, China, India, France, Israel and Netherland.

It is used in fresh and processed form. Among the vegetables maximum attempts have been made to improve this crop because of its short duration, easy cultivation and large number of seeds per fruit and this has made it an ideal crop for many research works. An exclusive Tomato Genetics Co-operative located in University of California, USA, published genetic information on the crop and maintains a large number of collections which have already been thoroughly catalogued.

Origin and Distribution: Native of Peru in South America and spread to North America by migrating birds. Largest number of wild tomato forms is present in Mexico. Spanish Priests introduced it to Europe in around 1550. In Europe it is known as Poma amoris = Amorous apple (Love apple). Wild relatives found in narrow elongated mountainous regions of Andes in Peru, Ecuador and Bolivia and Galapagos islands. Domestication and cultivation of tomato appears to have first occurred outside its centre or origin. Domestication of tomato was first done by Mexican people i.e. Red Indians, which is considered as secondary center of origin. Now cultivated tomato is common in Peru but recently added to their diet.

Quite contrary is true in Mexico where great diversity is seen and it is evident that cultivars are being grown and name tomato comes from a word in the Nahuatl language, *tomatl*. And also from the Aztecs of Central Mexico who called it 'xitomatl' meaning plump thing with a navel. The specific name, *lycopersicum*, means "wolf-peach", it is also known by various names like 'Pomme damour' (French) means Love apple, 'Pomidoro' (Douch) means Golden apple. First written document an arrival of tomato in the old world appeared in 1554 by the Italian herbalist Pier Andrea Mattioli. It was also known as Poma Peruviana i.e. apple of Peru. First cultivar introduced to Europe probably originated from mexico rather than South America and these presumably yellow (Golden apple) rather than red in colour. Tomato entered North America only from old world through colonist. In Britain John Gerard believed that tomato is poisonous (stem, leaves, but fruit is safe) as it contains glycol alkaloids. On hot day of August 1820 an ordinary farmer Robert Gibbon, Johansson first ate tomato to demonstrate its edibility and then it spreads throughout the world as vegetable crop. Only after 1830 tomato gained popularity as a component of food. Recently, as late as 1880, British finally conceded that, tomato is edible.

Linnaeus called it *Solanum lycopersicum* Miller: *Lycopersicon esculentum*

Now restored as *Solanum lycopersicum*

Lykos = wolf, *persicon* = peach

Chromosome number: True diploid, 2n = 24,

When Lycopersicon was having genus status (earlier), it was divided into two sub genera:

1. *Eulycopersicon*: red fruited type, contains two species, called as *Esculentum* complex *L. pimpinellifolium* - current tomato, small fruited, self pollinated, few are cross pollinated. *L. esculentum*- common tomato, strictly self pollinated.
2. *Eriopersicon*: green fruited type (even after ripening remain green), called as *Peruvianum* complex.: *L. peruvianum* wild
   L. hirsutum- wild, self fertile and self incompatible.
   L. chilense L. minutum wild, *L. glandulosum* wild,
   L. cheesmanii (wild, self pollinated).

Composition of Classification of Solanum genera: Adopted from Maria Jose Diez and Fernando Nuez, 2011 (Ed. Jaime Prohens and Ferando Neuz) who modified from Peralta and Spooner, 2006.

Miller, 1940	Luckwill, 1943	Rick, 1979	Child, 1990	Peralta, Knapp and Spooner, 2006
			Section Lycopersicon	
			Subsection Lycopersicon	**Section Lycopersicon**
Subgenus Eulycopersicon	**Subgenus Eulycopersicon**	**Esculentum complex**	**Series Lycopersicon**	**Lycopersicon group**
L. esculentum	*L. esculentum*	*L. esculennum*	*S. lycopersicum*	*S.lycopersicum*
L. pimpinellifolium	*L. pimpinellifolium*	*L. pimpinellifolium*	*S. pimpinellifolium*	*S. pimpinellifolium*
		L. cheesmaniae	*S. cheesmaniae*	*S. cheesmaniae*
		L. pennelli		*S. galapagense*
		L. hirsutum	***Series Neolycopersicon***	***Neolycopersicon group***
		L. chmielewskii	*S. pennelli*	*S. pennelli*
		L. parviflorum	***Series Eriopersicon***	***Eriopersicon group***
		Peruvianum Complex	*S. habrochaites*	*S. habrochaites*
Subgenus Eriopersicon	***Subgenus Eriopersicon***	*L. chilense*	*S. chmielewskii*	*S. huaylasense*
			S. chilense	
L. peruvianum	*L. peruvianum*	*L. peruvianum*		*Coraetiomulleri*
	L. pissisi?		*S. peruvianum*	*S. peruvianum*
				S. chilense
L. cheesmaniae	*L. cheesmaniae*		*S. neoricki*	***Arcanum group***
L. hirsutum	*L. hirsutum*			*S. arcanum*
L. glandulosum	*L. glandulosum*			*S. chmielewikii*
				S. neorickii

Tomato - Profile

Scientific Name	*Solanum lycopersicum*
Family	Solanaceae
Chr. Number	True diploid, 2n=24
Origin	Peru in South America
Breeding system	Self-pollinated
Anthesis	9 to 10 A.M.
Dehiscence	9 to 11 A.M.
Pollen fertility	On the day of Anthesis
Stigma receptivity	16 hrs before to 1 day after Anthesis

Evolution of Tomato

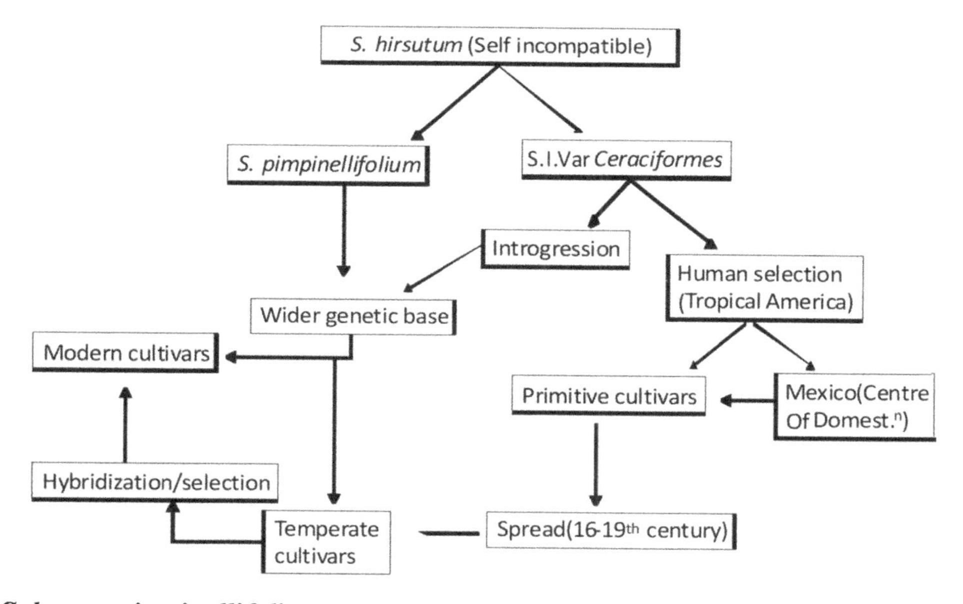

Solanum pimpinellifolium

The "Current tomato" in the other domesticated, edible tomato species. Not widely cultivated.

Solanum pennellii

This species has sticky hairs on the leaves and petals. You can see one function of these to catch pesky insects. The fruits are also fuzzy.

Survey of inter and intra specific breeding barrier in Lycopersicon (changed to Solanum)

F/M	*Solanum lycopersicum*	*Solanum pimpinellifolium*	*Solanum hirsutum*	*Solanum chilense*	*Solanum peruvianum*
Solanum lycopersicum	+	+	+	E.A	E.A
Solanum pimpinellifolium	+	+	+	E.A	E.A
Solanum hirsutum	+, U.I	+ , U.I	+, S.I	?	E.A
Solanum chilense	U.I	U.I	?	S.I	E.A
Solanum peruvianum	U.I	U.I	U.I	E.A	S.I

U.I: Unilateral incompatibility, S. I. : Self incompatibility, E. A.: Embryo abortion, +: good seed set.

Reproduction Biology

Tomato is one of the crops where genetic studies were conducted extensively because of wealth of variability within the species and ease with which it can be manipulated. It is a highly self pollinated crop. Natural Cross Pollination varies from 0.5 to 4 %. Much higher rate of cross-pollination observed at Peru is presumably because of native insect vectors. Introduction to Europe resulted in change of stigma position from out-side to within the anther cone and led to reduction in per cent cross pollination. Tomato flower is perfect; 4-8 flowers in a cluster, for a compound inflorescence and single plant can produce > 20 inflorescences in life cycle. Natural wind is sufficient to releases the pollen to pollinate the stigma (enclosed inside). In green house manual vibration of open flowers required for effective pollination and fruit set. Emasculation for the purpose of controlled pollination must be done approximately 1 day prior to anthesis or flower opening to avoid the accidental self-pollination. At this stage (1 day prior to anthesis) sepals began to separate and anther & corolla are beginning to change from light to dark yellow, the stigma appear to be fully receptive at this stage and can be pollinated under normal favourable conditions >200 seeds can be obtained from a single pollination. Protect the pollinated flower by covering it with butter paper bags. Cool, dry & relatively wind free weather is preferred for high success rates with outdoor crossing. Under optimal temp. and growth conditions tomato will complete its reproductive cycle in 95-

115 days, depending upon the cultivar. First flower opening take place 7-8 weeks after seeding, takes 6-8 weeks from first flower to ripe fruit. Under green house conditions thus makes it possible to complete 3 life cycles in year. Self-incompatibility (SI) is a common feature in wild relatives of the tomato and is transmitted to hybrids with *S. lycopersicum*. SI is conditioned by a single locus. Genic male sterility reported frequently within the genus and many loci producing ms genes have been identified and described.

Genetic Studies

Ideal plant for genetic studies because of its simple reproductive biology, ease of culture and wealth of genetic variations in cultivated and wild forms. Except maize it has been more extensively studied genetically than the any other major food crop. More than 970 genes reported in 1970. **'Tomato Genetics Cooperative'** established in 1951 by Dr. C.M. Rick, Univ. of California, Davis giving wide service to tomato workers and publishing descriptive list of new genes periodically. Genetic map developed showing the relative location of many genes controlling a wide variety of traits and is useful in designing & planning of breeding programmes, since linkage distance can be used to predict the probability of recombination between linked genes.

Breeding History

Livingstone in 1870 is generally recognized as the first tomato breeder in the North America. Early improvement credit should go to those who first domesticated, cultivated and consumed the crop. Many hundred of new cultivars developed within the past 50 years to meet the diverse needs and varied situations & climates. Machine harvestable tomato cultivars with small vine, concentrated fruit set and adequate firmness to withstand machine handling with high yield, quality and resistance. Taiwan was one of the leading producer of hybrid tomato seeds. Spontaneous mutation in Florida in 1914 used to develop the several varieties.

Genetic Resources

sp self pruning-det., few nodes between the inflorescence and early.

j-1 jointless pedicel- **Penn Red**

j-2 jointless pedicel- Many cultivars

I-1 Fusarium wilt resistance (against race-1) in **Pan America (*L. pimpinelli-folium*)**

I-2 Fusarium wilt resistance- **Walter**

Fusarium wilt resistance (race-1&2) **PI-126915**

I-3 Against race-3- ***L. pimpinellifolium* & *L. esculentum* lines US 629 & 638**.

Mi Nematode resistance-NTDR-1 from ***L. peruvianum*** (also source of leaf curl virus resistance & high ascorbic acid content)

L. peruvianum : possess resistance for Verticillium wilt, leaf curl and high TSS (10-11%).

L. pimpinnellifolium: possess resistance for Verticillium wilt, fusarium wilt, leaf curl and rich in 'Vit C and carotene content.

L. hirsutum-Insect resistance (colarodo potato beetle, Aphids & fruit borer), 2-tridecanone chemical governs the resistance.

L. hirsutum f.glabratum: possess resistance for leaf curl, Early blight, possess resistance for insect & low temp. tolerant.

L. cheesmanii : TSS (16%) & Salinity tolerance.

L. pennelli: Drought and salinity tolerance.

L. chemielentiskii: small bushy plants source of septoria leaf spot resistance.

Genetics: Qualitative Characters

1. Free neck (a weak peduncle – pedicel joint)
   Single recessive gene 'fn' (fn+)
2. Locule number: low = single recessive 'lc' (lc+)
3. Ovate fruit : ovate = single recessive 'o' (o+)
   Tight linkage between 'fn' and 'o'– Closely placed
4. 'B' high β-carotene; low lycopene
5. c – Potato leaf
6. Cl1 – Cleistogamy
7. j – Joint less pedicels
8. r – Yellow colour of fruit flesh
9. sp – Self pruning-determinate Plant habit
10. u – Without dark shoulder

Gene Designation	Gene Symbol	Source
Dwarf	d	Epoch & Tiny Tim
Potato Leaf	c	Geneva II
Verticillium wilt resistance	Ve	VR Moscow
Septoria resistance	Se	Targinnie Red
Late blight resistance	Ph-1	New Yorker
Alternaria resistance	Al	South Land
TMV resisrance	Tm-1,Tm-2, Tm-2^z	-
Spotted wilt virus	Several genes race specific	Many
Uniform ripening	u	Heinz -1350
High pigment	hp	Redbush
Green stripe	gs	Tigerella
High B-cerotene	B	Caro-rich
Parthinocarpic fruits	Pat-2	Severianin

Genetic Studies

Genetic variability, heritability and genetic advance: ***High genotypic variations recorded for:*** No. of fruits/plants & yield/plant (Singh *et al*., 1974). Wt. of fruit & early yield/plant (Dudi *et al*., 1983). Titrable acidity, pH, no. of locule/fruit, fruit set, seed content & seed germination.

High heritability and high genetic advance for: Yield, no. of fruits, no. of locule/fruit& fruit width. (Singh *et al*., 1974). Amino acid, reducing sugars & dry matter content. (Bhutani *et al*., 1974). No. of seeds, no. of fruits & juice/ pulp ratio. (Bhutani *et al*., 1974).

High heritability and low genetic advance for: Fruit diameter (Singh *et al*., 1988) Indicating involvement of additive and dominant effect of genes.

Correlation Studies

No. of fruits positively correlated with yield. Total yield positively correlated with early yield, number of fruits/plant, fruit wt., NPB, NSB & no. of fruits/ cluster.(Anad *et al*., 1986). TSS positively correlated with acidity & TSS/acidity ratio.

TSS : Acid ratio - acidity & no. of locules.

Path analysis: No. of fruit/plant, Green plant weight, Early yield /plant & Fruit weight. have high direct effect on yield.

Combining ability: High gca for yield, early yield, fruit size and fruit no. & *sca* for total yield, early yield and fruit no. depending on variations involved in the character.

Gene Action: Quantitative Charaters

Total yield – Non additive

No. of fruits – Additive

Early yield & fruit wt. – Additive

Plant height – Additive, non additive, additive and non additive different respects

Days to harvest : Additive

Locules per fruit 1. Additive and non additive

TSS, Acidity, Amino acids – Additive and non additive

Plant : Non- additive

Lycopene : 1. Non additive 2. Dominance

Ascorbic acid, Reducing sugars – Non Additive

Yield – Additive , Non additive and Additive and dominant

Trait Profile

1. **Earliness:** Direct seeding has enhanced the importance of earliness. Earliness can be observed in terms of :

   i) Time from planting to flowering

   ii) Time from flowering to initiation of ripening.

   iii) Concentration of flowering (no. of flowers produced per unit time).

   Earliest cultivars under optimal conditions will produce mature fruits in less than 90 days after seeding. Pusa Ruby (60-65 DAT), Pusa Early Dwarf (55-60 DAT), Arka Shreshta (50-60 DAT).

2. **Growth habit:** The gene 'sp' gene is responsible for determinate growth habit. This recessive gene gives smaller vine size, and inflorescence at closer nodes (<every 3rd node). The other genes are for dwarf (d) & brachytic habit (br/bk) (internodes greater than normal).

   All these genes can be used for Very compact growth habit, High population per unit area and for Machine harvesting.

   Determinate cultivars: HS101, Sweet 72, PED, Punjab Chhuhara, Pusa Gaurav, KS-2, CO-3, Arka Alok and Arka Ashish.

   Semi determinate: Sel.-120, Arka Shreshta, Arka Abhijeeth, Arka Vikas, Arka Saurabh, Arka Abha, Arka Ahuti, Arka Meghali.

   Indeterminate: Punjab Tropics, Pusa Ruby, T-1, Pant Bahar, Improved Meeruti, Arka Vardhaan, Arka Vishal.

3. **Mechanical harvesting:** Genotypes should have compact growth habit, concentrated fruit set, vine storability (fruit storability on vine after ripening), withstand mechanical harvesting jerk. Joint-less genes useful for mechanical harvesting. It is difficult in **humid areas (compared to arid zone). It requires a genotype to possess r**esistance to rain induced fruit cracking, tolerant to major fruit rots and more concentrated fruit set.
4. **Fruit quality:** Mature tomato fruits contain 94-95 % water, 5-6 % organic constituents which includes sugars 55 % (fructose, glucose, sucrose), alcohol insoluble solids 21 % (protein, cellulose, pectins, poly saccharides), organic acids 12 % (citric, malic, galacturonic, pyrrolidone, carboxylic), 7 % inorganic compounds and miscellaneous 5 % (carotenoids, ascorbic acid, amino acids, volatile compounds etc.).

Fresh Market

Quality Parameters

Appearance: Parameters like fruit size, shape, external colour, smoothness, uniformity, freedom from defects (resistance to fruit cracking), and fruit firmness are important.

Fruit colour: 'og^c 'gene- crimson - increases lycopene at the expense of β-carotene and reduces vitamin 'A'. 'hp' gene-high pigment - increases total fruit carotenoids, but exhibits pleiotropic effects (undesirable traits) like slow germination, slow growth and premature defoliation. Its use is limited in tomato improvement. Visually fruit colour can be measured by calorimetric method or Hunter lab colour difference meter can be used to assess the components of colour of juice (redness –a; yellowness –b; lightness –c).

Texture and firmness: Texture (smoothness), firmness and ratio of fruit wall to locular contents are the important quality parameters at the expense of consumer preferences for solving problem of long distance transport and storage.

Flavour: Sugars and organic acids determine the flavor of tomato. Proper balance of these constituents i.e. sugar: acid ratio gives optimal flavour. Relative levels of each (sugar & acid) gives intensity of flavor i.e. sweetness or sourness. Volatile constituents also contribute to tomato flavour.

Nutritional value: Vitamin 'A' and vitamin 'C' are the major vitamins present in the tomato fruit. Vitamin'A': Oxidation of one molecule of β-carotene (orange pigment) yields two molecules of vitamin 'A'. compounds that are converted in vivo to vitamin 'A' are termed as pro-vitamin 'A'. Certain carotenoids in tomato fruit alone may be converted to vit. A, but lycopene, the major pigment has no pro-vitamin 'A' activity. Orange fruited varieties have high vitamin 'A' activity than red fruited varieties. A single dominant gene 'B' favour β-carotene synthesis at the expense of lycopene and results in orange fruit with pro-vitamin 'A' level 8-10 times higher than the red fruited cultivars. The gene 'hp' increases pro-vitamin 'A', 'og^c 'gene decreases pro-vitamin 'A'and increases lycopene which is a potential antioxidant. Consumer preference is for red fruit. Vitamin'C': Ascorbic acid content of the fruit of *S. lycopersicum* varies from 10-20 mg/ 100g fresh weight. Linkage or pleiotrophy between high ascorbic acid content and small fruit size limited the use. The gene 'hp' also enhanced ascorbic acid content.

Processing quality: Uniform ripening or maturity state is important for evaluation of lines for processing quality.

Colour: It is a measure of fruit maturity and it influences the grades and standards of the processed commodity. Colorimetric measurement of raw juice is a standard practice.

Fruit pH: It affects heating time required to achieve sterilization of the processed commodity: high pH longer time is required. Values above pH 4.5 are unacceptable for fruit destined for concentrated products in which sterilization is achieved by pre-progrmmed heating times. For assessing lines, pH is measured directly with fresh juice prepared from a uniform sample of fully ripe fruits.

Over mature fruit will give erroneously high pH values.

Titrable acidity: It provides measure of organic acids (total acidity) which in terms estimates tartness. Total acidity and pH are not always closely correlated due to difference in the degree of buffering of pH by other fruit constituents. 10ml of fresh raw juice is diluted to 50 ml by adding distilled water. The volume of 0.1 NaOH required for titration to pH 8.1 is multiplied by correction factor (0.064) to estimate titrable acids as percentage of citric acid.

Soluble solids influence flavor and degree of concentration required (solid content). High total soluble solids lead to high recovery and require less energy in attaining desirable concentration of the processed product. Refractometer is used to measure total soluble solids. Total solids are influenced by environment and genetic factors. High light intensity, long photoperiod, dry weather increases TSS. Therefore selection for high yield or compact growth habit frequently results in low solid content. Total solids comprises of solid fraction (free sugars, organic acids) and insoluble fraction (proteins, pectins, cellulose and polysachharides). The insoluble fraction contributes for viscosity (consistency) of processed products, the most important component contributing for viscosity is polygalacturonides.

Viscosity: Important for estimating grades and standards of juice, ketchup, sauces, soup and paste. Acid efflux method or viscometer is used to estimate the viscosity.

Canned products: Whole tomatoes, uniform colour, size, shape, small core size and joint less pedicel (j-2) are essential to achieve good quality.

Fruit ripening: Ripening inhibitor (rin), non ripening (nor), inhibit the changes that accompany ripening, including colour development, fruit softening, ethylene production and respiratory changes. Their effects on enhancing shelf life in homozygotes or heterozygotes have prompted interest in their use in applied breeding programmes, particularly fresh tomatoes destined for long distance shipping. E.g.: Shelf life: FLAVR- SAVR is developed through genetic engineering by using anti-sense RNA technology. Processing varieties: Roma, Punjab Chhuhara, Arka Ashish, Arka Ahuti. Juice and Canning: Roma, San Marzano, Red Top, HS110. Puree: Indo Processor II, Rupal, Indo Processor III, Paste and Ketchup: HS101 and HS110.

Germplasm Resources

Germplasm maintained at 1. USDA – USA 2. AVRDC – Thaiwan. 3. Tomato Genetic Co Operative – Univ. of California, USA.

India : IIHR, Bangalore, Indian Institute of Vegetable Research, Varanasi.

IARI, New Delhi, NBPGR and different State Agricultural Universities.

Variability in Fruit Juice and Juice Characters

Growth types : Determinate, Semideterminate and indeterminate.

Fruit shape index = 0.7 (Pusa Ruby) to 1.43 (Italian Red Pear)

Pericarp thickness (cm) – 0.22 cm (Sabour Prabha) to 0.70 cm (IIHR 674 SB)

Juice yield (%) : 70.55% (HS 100) to 91% (Prista)

TSS (%) : 4.2% (Preslav) to 6.00 (Pusa Ruby)

Pulp content (%) : 21% (Preslav) to 30.40% (Pusa Ruby)

Lycopene (mg/100 g) : 1.57 (EC 154896) to 6.77 (Pusa Ruby)

Beta carotene (mg/100 g) 12.20 (Patriot) to 60 (Indo Process II)

Ascorbic acid (mg/100 g) : 10.44 (Karnataka Hybrid) to 22.10 (AC 238).

Breeding Objectives: Objectives vary with purpose e.g. Processing, fresh market, greenhouse production, home gardening

General Objectives

Early and high yielding varieties / hybrids

Resistance to important pests and diseases of the area (virus, bacteria)

Resistant to root knot nematode

Suited to cold and high temperature

Suited to prolonged storage and transportation

Suited for processing

Multiple insect and disease resistance

Quality traits-free from the defects, appearance, cracking, marketability, nutritional etc.

Resistance to salt, drought etc.

Breeding Programs

Hybridization followed by pedigree selection has been the most commonly used breeding method for tomato improvement. Backcross breeding has been the method of choice in wide crosses or interspecific gene transfer. In certain situations, a combination of pedigree selection and back cross breeding has proven useful to exploit the advantages of each method. With offseason breeding nurseries, tomato breeding in west has become a year round activity with two or three generations each year. SSD has been examined as an alternative to pedigree selection for use where facilities or funds do not permit maintenance of winter breeding nurseries. SSD was most efficient in both time and progress under selection (Casali and Tigchelaar, 1975). However efficiency depends on breeding objectives and heritability. SSD is followed for characters of low heritability, since a broader genetic base is maintained to advanced generations. Pedigree selection is most effective with high heritability. Selection for several characters simultaneously a combination of two methods appears desirable. Pedigree selection is practiced in early generations (F2 & F3) for highly heritable traits followed by selection among SSD - derived inbreds F6 or F7 lines for characters of lower heritability.

Generation	Area/place	No. of plants
F1	Area of intended use	10-15
F2(Pedigree selection)	Winter nursery	250-300
F3(Pedigree selection)	Area of intended use	125-150
F4 (SSD)	Green house	15-20 plants for each F3 selection
F5 (SSD)	Green house	15-20 plants for each F3 selection
F6 (SSD)	Area of intended use	15-20 plants for each F6 SSD line

Donors for Desirable Traits in Tomato

Fusarium wilt : Pan American, *S. pimpinellifolium*, *S. hirsutum*, PI 13448, *S. peruvianum*, EC-148898, Roma, Colombia, HS 110

Verticillium wilt : VR Moscow,

Late blight: West Virginia 63

Early blight : 68B134, Southland, *S. pimpinellifolium*

Anthracnose : PI 272636,

Bacterial wilt, PI127805A, Saturn, Venus

Bacterial canker : Bulgarian 12, Utah 737

Tomato mosaic : Ohi M29

Leaf curl virus : *S perivianum* and *S. hirsutum*

Fruit warm: PI 126449 (*S. hirsutum*)

Whitefly : IVT 74453 (*S. hirsutum*)

Cold tolerance : *L. pimpinellifolium*, Fire ball, Red Kloud, Girld, Oregon-11, Immuna, Cold Set, Pusa Sheetal, Avalanche, Tempo, Out door,.

Heat tolerance : Phillippine, Punjab Tropic, *S. cheesmanii,* EC 130042, EC 162935, PS1

Salt tolerance : *S. cheesmanii*, Sabour Suphala

Drought tolerance : *S. pennelli,* IIHR 14-1, 146-2, 383, 553, 555

High ascorbic acid : *S. Pimpinnellifolium*, Double Rich

High TSS : A 276, A408, Red Cherry

Red colour : High pigment, Crimson

B-carotene : Caro Red, Caro Rich

Breeding Methods

1. **Introduction** : Sioux, Marglobe, etc.
2. **Pure Line Selection** : Tomato self pollinated crop
   Arka Vikas: suited for rainfed, Pure Line Selection from Tip-Top (USA)
   Arka Saurabha: PLS from V-685 (Canadian Breeding line)
   Arka Abha: possess resistance for bacterial wilt, PLS from VC-8-1-2-1 (from AVRDC, Thaiwan)
   Arka Alok: possess resistance for bacterial wilt, PLS from CL-114-5-1-0 (from AVRDC, Thaiwan)
   Arka Ahuti: possess resistance for processing, PLS from Ottawa-60 from Canada

3. **Single seed descent method** : Suited for characters with low heritability (yield and yield components, earliness) and in crops with wider spacing.
4. **Mass selection** : Massing of IHR 674 from UC82-line from USA developed in to Arka Ashish: processing variety with resistance to powderymildew.

   Mass selection is not extensively used.
5. **Mutation breeding** : S12 from Sioux

   Pusa Lal Meeruti from Improved Meeruti (seeds irradiated with 15-30 Kr g-rays).

   Natural mutants: 'Nr'- never ripening, 'rin' ripening inhibitor, 'nor'- non ripening, 'alc'- slow ripening.
6. **Pedigree method** : Two parents selected based on wide genetic divergence and desirable traits.

   Crosses are made and selection starts from F2.

   It is possible to improve characters governed by additive and non additive genes.

   Eg: Pusa Ruby selection from cross between Improved Meeruti x sioux.

   Arka Meghali: possess resistance for drought – Pedigree selection (F8) of the cross Arka Vikas x IIHR-554.
7. **Back cross method to transfer characters between species: It is easier when few genes are involved**. Bacterial wilt resistance, early blight resistance and nematode resistance breeding programs it can be employed.
8. **Heterosis breeding :** Tomato is a classical example in self pollinated vegetables where heterosis is being exploited on commercial scale. Hendrick and Booth (1907) were the first to report on the presence of heterosis in tomato. Heterosis is reported for earliness, plant height, number of branches, leaf area, fruit size, locule number, fruit number, fruit yield, chlorophyll content, ascorbic acid, pH, TSS etc.

Heterosis: it is easy to develop F1 hybrids with earliness, large fruit size, high yield and adaptability compared to pure line development. Private Companies developed such hybrids with hand emasculation and pollination technique for F1 seed production and hybrids for quality parameters like amino acids, juice yield, carotenoids and lycopene. There is a vital scope for F1 hybrids resistant to diseases like LC, FW, BACTERIAL WILT, EB and LB. Tm-2 and Mi gene has undesirable effects at homozygous state and best exploited at heterozygous state. Therefore LC resistance and nematode resistance can be exploited through heterosis breeding.

Methods of Hybrid Seed Production

1. Hand emasculation and pollination

The ratio of female to male parents is 12:1.

One kg of hybrid seed can be produced by 40-45 persons in one day.

Flowers of seed parents are taken for emasculation at a stage of 12-15hr before opening and anthers are removed by forceps, needle or finger nail etc. Generally emasculation is done in afternoon and pollination is done in the following morning. After emasculation anthers are collected and overnight kept under light and morning pollens are extracted. Pollen is mixed with talcum powder filled in finger cups and used for pollinating the emasculated flower and immediately after pollination two sepals are removed as a marking. Later such marked fruits only harvested as a source for F1 seeds. Pollen can be stored for 2-3 days under normal condition but fruit set reduces. With fresh pollen, hand pollination results into 90% fruit set and depends upon temperature, soil moisture etc.

Released F_1 hybrids

Arka Vishal	:	IIHR837 X IIHR 932 – ID (140 g).
Arka Vardan	:	IIHR550-3 X IIHR932 – ID, possess resistance for Nematodes – 140 g.
Arka Shreshta	:	15SBSB X IIHR1614 – SD, possess resistance for bacterial wilt (70-75 g).
Arka Abhijit	:	15SBSB X IIHR1334 – SD, possess resistance for bacterial wilt (65-70 g).

2. Use of male sterile lines

Male sterile lines can be identified by small flower size, poorly developed anthers, poor or no anther dehiscence, absence of pollen grains, sterile pollens.

Genic male sterility can be maintained at heterozygous state and markers used to identify male sterile plants are Potato leaf, Anthocyanin less (aa) and brown seeded.

Genic male sterility MS series with double tagging have been developed to overcome the segregation of sterile and fertile plants at the time of seed production.

Use of *CMS* lines: Use limited due to pollination problem.

3. Use of functional male sterility

Also called as John Bay type male sterility. Anther do not dehisce. Closed anther types. Developed hybrids: Katrain Hybrid No.I, Katrain Hybrid No. 2. Line 510 is a functional male sterile, 'clm' a single recessive gene controls it. Other hybrids – Meeruti clm 2 x sioux, Meeruti x Red jacket.

4. Use of structural male sterily (Stamenless type)

Stamenless controlled by monogenic recessive but not exploited.

5. Use of exerted stigma type as female parent. Eg. Manaleuci X Bruinsma (exerted stigma line).

6. Use of selective differential gametocides

Like FW-450, 24D, MH, Causes pollen sterility. MH 100 ppm once or 50 ppm twice, Female fertility was not diminished.

In India Majority hybrids seeds are produced by hand pollination and emasculation, Examples:

Determinate group: ARTH3, Pusa Hybrid 2, NARF 101, Swarna12, Nath Amruth 501, NDTH6, HOE 303, HOE 606 and DTH 4.

Indeterminate group: ARTH4, FM2, KT4, MTH6, FM1, Nath Amruth 601 and NDTH2 evaluated during 1990-91 under PDVR.

9. Interpsecific hybridization

1. *S. peruvianum:* Resistant to *Verticillium* wilt, ToLCV, nematode and salinity
2. *S. pimpinnellifolium*: Resistant to Bacterial wilt and *fusarium* wilt and rich in Vititamin-C and high β-carotene content
3. *S. hirsutum*: Fusarium wilt, insect resistance
4. *S. cheesmani* : Salinity resistance
5. *S. pennelli* : Drought and salinity resistance

Eriopersicon group difficult to cross with Eulycopersicon group and even if fertile, in F1 sterility and SI is seen. Embryo rescue method and following techniques can be adopted to overcome cross incompatibility.

1. Mixed pollination.
2. Grafting and pollination – style and stigma.
3. Use of IAA and pollination.
4. Use of mutated pollen.

10. Polyploid breeding

Autotetraploids have been produced through colchicine treatment (0.2 per cent), but are of no economic significance. Tetraploids are sterile and fruit size is small.

11. Genetic engineering / Biotechnology

1. Antisense *RNA* for PGU used and developed FLAVR SAVR variety with long shelf life.
2. Viral coat protein: virus resistance
3. Herbicide resistance: weed control
4. Insect resistance: Bt gene

Breeding for Storage Life and Processing

Processing type are distinct from fresh market types for high TSS, high acidity, uniform intense red colour, resistance to cracking and elongated fruits with high yield. Plants should be adapted to mechanical harvesting with determinate habit, concentrated fruit set for once over harvest and jointless pedicel of fruit for easy seperation. Eg. Roma, Punjab Chhuhara, Arka Ahuti and Arka Ashish. For juice or canning of whole peet tomato : Roma, San Marzano, Red Top, HS 110.

Puree preparation : Indo-Process II, Indo Process III and Rupali

Paste and ketchup : HS 101 and HS 110

Resistance Breeding

Buck eye rot: ***Phytopthora parasitica***

It is a problem in Himachal Pradesh. Resistant varieties are San Mazano, Molokai, Red Cherry. Resistance to BER in *Solanum pimpinellifolium* var. EC54725 was controlled by a dominant gene 'Br' (Rattan and Saini, 1979).

Septoria leaf spot: *Septoria lycopersicii*

Resistant sources are: KT4, EC2750, EC7785, Targinnine Red.

Fusarium wilt: *Fusarium oxysporum fsp. Lycopersici*

Pan America is resistant to Race 1 which is transferred to L-93, L-94, L-203 (Porte and Walker, 1941). Walter is resistant to Race 2 and Race 1 (Strobel *et al.*,1969). *L. hirsutum fsp glabratum, L. hirsutum* PI13448, *L. peruvianum* EC148898 are immune. *L. esculentum* cv. Columbia, Roma, HS110 are resistant. In India: Marglobe, Rutgers, Pant Bahar (Field conditions). Inoculation techniques: Dip roots in inoculums and transplanted to field.

Early blight: *Alternaria solani*

It is a serious problem in Punjab, Haryana and Maharashtra.

Resistant sources: 68B134, Kalyanpur Sel.1, Manalee.

68B134: inoculate the foliage with swarm spore suspension by atomization (Barksdale and Stoner, 1973).

Late blight: *Phytopthora infestans*

It is a serious problem in H.P, U.P. *L. pimpinellifolium* is a source of resistance.

Sources is West Virginia63 and the resistance is transferred to L-1197, L-1497. Inoculate the source (West Virginia63) leaves with swarm spore suspension by automization (Gallegly, 1960).

Bacterial wilt: *Ralstonia solanacearum*

Serious in Karnataka, Kerala, Bihar, M.P, Orissa. Source are PI127805A, Saturn, venus. Later the resistance is transferred to L-1, L-8, L-15, L-21, L-29 etc.

Other sources are Acc 99, Sweet 72, Acc 151, IIHR 663-12-3, Arka Abha, Arka Alok, Megha. RT-L is released at national level.

F1 hybrids resistant are Arka Abhijit, Arka Shreshta.

Inoculation technique: Cut roots on one side and pour bacterial suspension into soil trench (Henderson and Jenkins, 1972).

Many isolates from Race-1 reported. Race-3 infects potato and tomato and varies in pathogenesis.

S. pimpinellifollium (PI 127805-A) carry resistance which has partial dominance.

VC-8-1-2-1 (AVRDC Taiwan) source to Arka Abha and CL-114-5-120 (AVRDC Taiwan) source to Arka Alok where, resistance is controlled by single dominant gene. Resistance varied with isolates and disease resistance reaction varies from variety to variety and isolate to isolate. In line LE79 resistance is governed by monogenically incomplete dominance (USA).

Powdery mildew: Arka Ashish is a source of resistance.

Verticillium wilt: *Verticillium albo-atrum.* Source of resistance is VR Moscow.

Inoculation technique: Dip roots in the inoculums and transplant to field (Walter, 1967).

Bacterial canker: *Corynebacterium michiganensis:* Source is Bulgarian 12, Utah 737.

Bacterial speck: *Pseudomonas solanacearum:* Source is Ontario 7710.

Grey leaf spot: *Stemphyllium solani:* Source is Manalucie

Leaf curl / Yellow leaf curl: caused by Gemini virus and transmitted by *Bemisia tabaci*. Source is LA-121 (*S. pimpinellifolium*). Screening technique: Release viruliferous whitefly female on to caged tomato seedlings and allow feeding for 72 hours (Pilowski and cohen, 1974).

Other sources: *S. hirsutum fsp. glabratum* where resistance is linked with small fruit size and late maturity. *S. peruvianum* (PI 127830 & PI 127831) and *S. pimpinellifolium* are also sources where resistance is linked with small fruit size and late maturity and controlled by single incomplete dominance. *S. hirsutum*: LA-386 & LA-1777, resistance is controlled by complementary factors. *S. lycopersicum*: tolerant lines are HS 101; HS 102, Nematex,T1 & Novel (but susceptible at Hissar conditions). Resistance is associated with increase in phenolic contents, non-reducing sugars and total sugars. Resistance to TCLV from *S. hirsutum fsp. glabratum* is transmitted to *S. lycopersicum* lines H-2, H-11, H-17, 23& 24 by using backcross method and also from *S. pimpinellifolium* to *S. lycopersicum* lines LCP-22, 2, 3, 9.

Spotted wilt virus: Source: *S. peruvianum* & *S. pimpinellifolium*, Pearl Harbour. Screening technique: place the seedlings in disease nursery and encourage the thrips (Frazier *et al.*, 1950).

Tomato mosaic: caused by TMV.

Source: Ohio M-R9. Apply the expressed inoculums with an air brush, inoculate again 10days later. Curly top:Source: CVF4 Release viruliferous leaf hopper into screened green house 2-3 times at 1week interval (Martin, 1969).
Genes carrying resistance to the important tomato diseases.

Insect Resistance

Fruit borer: polyphagous

Source: *S. hirsutum fsp. glabratum* PI 126449, from this resistance is transmitted to lines L-139, 177, 181.

Less infected accessions are 128, 133, 145, Heinz 137 & Sabour Prabha.

S. hirsutum fsp. glabratum highly resistant (Kashyap and Verma, 1986).

Genetics: Preponderance of additive gene effects, narrow sense heritability is high.

Biochemistry: Antibiotics and phenolic compounds adversely affect the life cycle of insect.

Biotechnology: ‘Bt’ gene transferred from *Bacillus thurungiensis* to cultivated types by pvt. Companies.

Mechanism of resistance: *S. hirsutum* contains a factor highly detrimental to development of fruit worm larva, resulting in high larval mortality (Fery and Cuthbert, 1975).

Potato aphid: Source: *S. peruvianum* PI 129145, *L. pennelli.*

Disease	Pathogen	Gene of resistance	Source	Reference of genetic control
Fungi				
Verticillium wilt	*Verticillium dahliae*	*Ve*	*S. pimpinellifolium*	Cannon and Waddoups, 1952
Fusarium wilt	*Fusarium oxysporum f.* sp. *lycopersici*			
	---pathotype 0	*1*	*S. pimpinellifolium*	Kesavan and Choudhuri, 1977
	---pathotype 1	*1-2*	*S. pimpinellifolium*	Alexander and Hoover, 1955
	---pathotype 2	*1-3*	*S. pennelli*	Scott and Jones, 1989
Alternaris stem canker	*Alternaria alternate f.* sp. *lycopersici*	*Asc*	*S. lycopersicum*	Clouse and Gilchrist,1987
Grey leaf spot	*Stemphllium* spp.	*Sm*	*S. pimpinellifolium*	Andrus *et al.*, 1942
Leaf mould	Fulvia fulva *(Cladosporium fulvum)*	*Cf(1to24)*	*S. pimpinellifolium* *S. lycopersicoides* *S. habrochaites* *S. peruvianmum*	*Kerr et al., 1971*
Powdery mildew	*Leveillula taurica*	*Lv*	*S. chilense*	Stamova and Yordanov, 1990
	Oidium neolycopersici	*Ol-1* *Ol-2*	*S. habrochaites* *S. lycopersicum*	Van der Beek *et al.*, 1994
Late blight	*Phytophthora infestans*	*Ph-1*	*S. pimpinellifolium*	Pierce, 1971
		Ph-2	*S. pimpinellifolium*	Moreau *et al.*, 1998
		Ph-3	*S. pimpinellifolium*	Chunworgse *et al.*, 1998
Fusarium crown and root rot	Fusarium *oxysporum* f.sp. *radicis lycopersici*	*Frl*	*S. peruviamum*	Berry and Oakes, 1987
Corky root	*Pyrenochaeta lycopersici*	*Pyl*	*S. peruvianum*	Laterrot, 1978
Viruses				
Tomato mosaic virus	Tomato mosaic virus (ToMV)	*Tm-1*	*S. hirsutum*	Pelham, 1966
		Tm-2	*S. peruvianum*	Laterrot and Pecaut, 1969
		$Tm\text{-}2^2$	*S. peruvianum*	Hall, 1980

Tomato spotted wilt virus	Tomato spotted wilt virus (TSWV)	*Sw-5*	*S. peruvianum*	Stevens *et al.*, 1995
Tomato yellow leaf curl virus	Tomato yellow leaf curl virus (TYLCV)	*Tylc* *Tv-1* *Ty-2*	*S. pimpinellifolium* *S. chilense* *S. habrochaites*	Kasrawi, 1989 Zamir *et al.*, 1994 Hanson *et al.*, 2000
Tomato leaf curl virus	Tomato leaf curl virus (TLCV)	*Tlc*	*S. pimpinellifolium*	Barenjee and Kalloo, 1987
Alfalfa mosaic virus	Alfalfa mosaic virus (AMY)	*Am*	*L. hirsutum f. glavratum*	Parrella *et al.*, 1998
Potato virus Y	Potato virus Y (PVY)	*pot-1*	*S. habrochaites*	Legnani *et al.*, 1995
Bacteria				
Bectrial speck	*Pesudomonas syringae pv. tomato*	*Pto*	*S. pimpinellifolium*	Pitblado and MacNeill, 1983
Becterial spot	*Xanthomonas compestris pv. vesicatoria*	*Bs-4*	*S. pennelli*	Ballvora *et al.*, 2001
Nematodes				
Root-knot nematode	*Meloidogyne incognita, M. arenaria*	*Mi, Mi-1* *Mi-3, Mi-9*	*S. peruvianum*	Smith, 1944
Potato cyst Nematode	*Globodera restochiensis*	*Hero*	*S. pimpinellifolium*	Ellis and Maxon-Smith, 1971

Leaf miner: Source: *S. hirsutum fsp. glabratum* PI 126449, *L. hirsutum* PI 126445.

Whitefly: Transmits leaf curl virus. Source: *S. hirsutum* IVT-74453, *L. pennelli* IVT-72100. No systematic studies but visual observation indicate *S.hirsutum* and *S.hirsutum glabratum* less affected by whitefly.

General mechanism of insect resistance: Entanglement of insects (aphids & thrips) in a gum like exudates from tomato foliage (Mc kinney, 1938), reduced oviposition due to high frequency of glandular hairs on the foliage (Hawaiian cultivars) of spider mites, flea beetle and whitefly and glandular hair secretion (ethanol washing reduce resistance). Increase in leaf trichome frequencies. Increase in level of **tridecanone** (a methyl ketone) secretion from leaf trichomes againt mites and tomato horn worm.

Root Knot Nematodes: *Meloidogyne incognata, M. arenaria, M. javanica*. Resistance to *M. incognata* was first reported in *S. peruvianum* PI 128657

(Bailey,1941) and through embryo culture resistant genes were transferred to *S. lycopersicum* (Smith,1944). Resistance is conditioned by a single dominant gene 'Mi' located on 6th chromosome, also provide resistance to *M. arenaria*, *M. javanica* and *M. acrita*. Several cultivars were developed by using the sources like Nematex and Resistant Bangalore are NRT-3, 8 & 12.

NRT-8 was named as Hissar Lalit.

Genes	Type	Variety
L Mi R1	Dominant	Nematex
L Mi R2	Dominant	Small Fry
L Mi r3	Recessive	Cold Set 1

Other resistant varieties are NTDR-1, Arka Vardhan, Pusa 120 & PNR-7.

Lines: L-97, 272, 274.

Mechanism of resistance to nematodes: Localized accumulation of **chlorogenic acid** and its oxidized products which inhibits nematode population increase.

Environmental Stress Resistance

Cold resistance: Seed germinaton at low temperature (<2- 4°C) is critical and growth and development should be good at low temperature which was observed in *S. hirsutum fsp. glabratum.* Fruit set is very critical at low temperature.

Sources: Cold Set, Montfaoct 63-4, Precoce and Severianin will set fruit at low temperature and more fruit set at very low temperature (0-2°C) observed.

Genetics: yield, no. of fruits and fruit wt. at low temperature is controlled by Additive and Additive Dominant components. Poor fruit set attributed to poor anther dehiscence at lower temperature. Parthenocarpic variety have advantage here. E.g.: Severianin a Russian variety where, parthenocarpy is controlled by a single recessive gene 'pat-2'.

Heat tolerance: Tomato in general can't tolerate more than 35°C. Critical stages include germination (no germination at higher temperature), growth, fruit set and ripening. Fruit set: line setting fruit at high temperature are Philliphine, Punjab Tropic, Mart-3 and P-4. *S. cheesmanii* accessions EC130042 and EC162935 exhibit high flowering even at high temperature. PS-1 set parthenocarpic fruit even at 40-42°C (controlled by partial dominance).

High temperature leads to poor fruit set due to stigma exertion, anther tip burning, poor anther dehiscence, low pollen viability, slow pollen tube growth and poor fertilization

But in resistant varieties fruit set is higher even at high temperature due to less stigma exertion, less anther tip burning, high pollen viability, better pollen tube growth and high fertilization. The lines identified for better fruit set at higher temperatures are having smaller fruit size that needs to be improved.

Drought tolerance: Drought tolerance is found in *S. chilense* (due to deep tap root system) and *S. pennelli* (limited root growth but has ability to conserve moisture). Screening at IIHR for drought tolerance was done and IIHR 14-1, IIHR 146-2, 383, 553 & 555 were found tolerant. At Hissar EC 130042 and Sel.28 are identified as tolerant lines.

S. *pimpinellifolium* accession PI 205009 & 84 # 88 plants remained green for long period. Sel.28, rin, 84 # 76 and JCL-4 gave more yield than other lines under drought conditions.

Drought Index: It is indicator of drought resistance and was high in *S. pimpinellifolium* EC 65992 (43%), Pan America (39%) and drought susceptibility check had low Drought Index of 4.4%

Drought resistant lines had increased no. of branches/plant, increased leaf area/plant, long root length, leaf rolling (drought avoidance, delays/postpones the stress), wax coating on leaves and increased Relative Water Content of leaf and decreased water saturation deficit, lower plant canopy temperature and increased osmotic potential.

From IIHR: Identified as Arka Vikas tolerant and Arka Meghali as resistant varieties.

Salt tolerance: germination stage is more critical.

Sabour Suphala, *S. cheesmanii* (most potential species), *S. hirsutum fsp. glabratum* exhibited germination at 8mmhos/cm (Lal *et al.*, 1981).

Mechanism

Osmotic: Less absorption of the salts leads to high water potential in the root tissue compared to nearby soils and there is moment of water from higher water potential to lower water potential zones as a result water stress is created.

Ionic: salt absorbed in large quantity leads to salt injury

Herbicidal resistance: Higher survival in Hissar Lalit was observed followed by Pusa Ruby, Sel.22, Sel.2 and Columbia against herbicides like Pendimethylene, Metribuzine, Butachlor, Oxadiazone and Alachlor under field conditions as reported by Bavarjee *et al.*, 1990.

Breeding Achievements

Tomato hybrids / varieties

Pusa Sadabahar : A determinate tomato variety developed by IARI, New Delhi, can set fruit up to 30^{0}c of night temperature and as low as 6°C, suited for North India, fruits ideal for distant market. Determinate, 40 g fruit size.

Kashi Hybrid-1 : Early for IIVR, 111 t/ha

H-86 (Kashi Vishesh): TLCV tolerant 48 t/ha from IIVR

Pusa Hybrid 8: Det. hybrid developed at IARI, ND, 60-70 g/fruit

Nandi, Sankranthi and Vaibhav are leaf curl resistant varieties developed at UAS, Bangalore.

Pusa Divya: New tomato hybrid by IARI Katrain by using close anther mutant with potato leaf marker in female parent i.e. Long Styled Psaiji X Roma as male parent, indeterminate, dark green stem end, matures in 70-90 DAT,35-45t/ha, firm fruit, round-oval, thick skinned, can be transported to distant places, less incidence of late blight and buck eye rot. Firm fruits suited for transportation. Indeterminate, 35-45 t/ha, 16 fruits weigh 1 kg.

Shakti (LE-79): KAU, Trissur, possess resistance for bacterial wilt.

Sonali: possess resistance for late blight.

TRB-1 & 2: PAU, Ludhiana, possess resistance for late blight.

Pant Bahar: possess resistance for late blight.

From OUAT, Bhubaneshwar:

Utkal Pallavi (Bt-1): possess resistance for bacterial wilt; **Utkal Deepti** (Bt-2): possess resistance for bacterial wilt

Utkal Kumari (Bt-10): possess resistance for bacterial wilt

Hissar Anmol (H-24): HAU, Hissar, possess resistance for leaf curl virus

Hissar Gaurav (H-36): HAU, Hissar, possess resistance for leaf curl virus

H-84: HAU, Hissar, possess resistance for leaf curl virus; H-88: HAU, Hissar, possess resistance for leaf curl virus.

H-86 (Kashi Vishesh): TLCV tolerant, 48t/ha.

Kashi Amrit: IIVR, Varanasi, possess resistance for leaf curl virus.

Avinash-2: Syngenta, possess resistance for leaf curl virus.

Tomato fruit borer resistant source: **Angurlata, Punjab Chhuhara, Sabour Prabha**.

Nematode resistant source: **Arka Vardhan, Hissar Lalit, Pusa Hybrid-2**.

Arka Ananya (TLBRH-9): A tomato hybrid (TLBR-6 x IIHR-2202) from IIHR, Bangalore is resistant to tomato leaf curl virus and bacterial wilt and is semi determinate and suited to both rainy and summer cultivation.

Pusa Hybrid 8: Tomato variety (60-70g fruits, 636.66q/ha) suitable to spring - summer cultivation was released from IARI, New Delhi.

NTDR-1 (UAS, Bangalore): resistant to nematode and good yielder.

Arka Ahuti: Medium sized, deep red colour fruits suited for processing and yields 45 t/ha.

Arka Ashish: Resistant to powdery mildew, fruits deep red with medium size, suited for processing, yields 40 t/ha.

Arka Meghali: Medium sized fruits, tolerant to moisture stress, yields 18 to 19 t/ha.

Arka Saurabh (IIHR): Semideterminate, fruits firm round, medium large, deep red and nipple tipped, yields 30 t/ha.

Arka Vikas (IIHR): Indeterminate, fruits medium to large, deep red colour, tolerant to moisture stress.

Punjab Tropic (PAU, 1971): Indeterminate, vigorous growth, fruits large, round, green stem end, less seedy, juicy suited for fresh market, matures late, yields 60 t/ha.

Pusa Ruby (IARI, 1975): Indeterminate, spreading, less branched and hardy. Fruits flat, round, small to medium (4-5 locules), slightly acidic, early maturing (60-65 days), yields 33 t/ha.

H.S. 101 (HAU, 1975): Determinate, multibranched, fruits round, small to medium, 3-4 locules, yields 25 t/ha.

S-12 (PUA, 1975): Determinate, bushy, fruits medium sized, compressed round, highly acidic and juicy. Suited for fresh market, yields 45 t/ha.

Pusa-120 or S-120 (IARI, 1975): semideterminate, spreading type, foliage dark green, fruits round to flattish round, medium to large in size, less acidic, resistant to nematodes, suitable for summer production.

T-1 (GBPUA&T, 1975): Indeterminate, fruits round, slightly pointed at the stigmatic end (4-5 locules) and non-cracking.

Megha or L-15 (UAS, Dharwad): Determinate, fruits small round, bacterial wilt resistant, suited for rainfed cultivation and yields 25 t/ha.

Sweet-72 (HAU, 1975): Determinate, fruits flattish round with green stem end.

Pusa Early Dwarf (IARI, 1975): Determinate, fruits flattish round, medium size, ribbed, early (55-60 days) and yields 39.5 t/ha.

Punjab Kesri (PAU, 1978): Determinate, fruits oval, Kesri round, firm, thick walled, trilocular, sour and seedy and juicy. Yields 70 t/ha.

Punjab Chhuhara (PAU, 1981): Determinate, fruits medium sized pear shaped, firm, fleshy, bilocular, less seedy, less sour, thick walled, good shelf life. Suited for long transport, yields 75 t/ha.

Pusa Gaurva or Sel. 152 (IARI, 1983): Determinate, fruits yellow, oblong, bilocular, ripens to uniform red colour, suited for processing and long transportation and yields 30 t/ha.

KS-2 (Kalianpur, 1985): Determinate, light green foliage, fruits flattish round, slightly furrowed with 4-5 locules.

Pant Bahar (Pantnagar, 1985): Indeterminate, light green foliage, stem thin, leaflets small, fruits flattish round, medium in size (5-6 locules), slightly ridged, resistant to verticillium and fusarium wilt.

Improved Meeruti (IARI): Indeterminate, medium maturity, fruits firm, flattened, grooved with distinct ribs, medium size, yields 32 t/ha.

CO_3 (TNAU): Determinate, fruits round, juicy and suitable for fresh consumption.

Punjab Nematode Resistant-7 (PAU, 1985): Determinate, fruits flattish round, medium to large, juicy and acidic, resistant to rootknot nematodes and fusarium wilt, yields 40 t/ha.

HS102 (HAU): Determinate, early bearing, fruits small, round, juicy, sets fruit at high temperature.

HS110 (HAU): Semideterminate, potato leaf type, late, fruits large, fleshy round with meaty texture, suited for table purpose.

Hisar Arun or Sel. 7 (HAU): Early bearer, fruits medium to large, round, yields 35 t/ha.

Hisar Lalit or NT 8 (HAU): It is resistant to root knot nematodes. Determinate, early, fruits round and medium sized.

Pant T3: Semideterminant, suited to winter season, fruits round, yield 20 t/ha, suited for processing.

There are many cultivars, *viz.*, Sioux, Best of All, Marglobe, Pusa Sheetal developed by IARI and Arka Vishal, Arka Vardan, Arka Abha, Arka Alok developed by IIHR.

OSU unveils new purple tomato, 'Indigo Rose': Indigo Rose, a truly purple tomato, from OSU's program to breed for high levels of antioxidants. (Photo by Tiffany Woods).

Arka Rakshak: High yielding F_1 hybrid with triple disease resistance to ToLCV, bacterial wilt and early blight, fruits are large (90-100g) and deep red, firm fruits suitable for fresh market and processing, yields 75-80 t/ha in 140 days. During 2012, the Indian Institute of Horticultural Research (IIHR) in Bangalore, introduced **'Arka Samrat' and 'Arka Rakshak,' two new triple disease resistant tomato hybrids.** AVRDC breeding lines **CLN-2498E (male parent of 'Arka Samrat') and CLN- 2498D (male parent of 'Arka Rakshak')** were used in the breeding program to create the new hybrids, which resist *Tomato leaf curl virus,* bacterial wilt, and early blight, three of the most serious problems in tomato production. AVRDC tomato breeder **Peter Hanson supplied the breeding** lines and worked with researchers at IIHR to stack multiple resistance traits into the hybrids.

Arka Samrat: High yielding F_1 hybrid with triple disease resistance to ToLCV, bacterial wilt and early blight. Fruits oblate to high round, large (90-110g), deep red, firm fruits suitable for fresh market. Yields 80-85 t/ha. in 140 days.

New entries like DT 39' ('Pusa Rohini'), 'DAG 1' ('Pusa Ujwal'), 'DTH-41, (Hybrids) 'TH 317' (Hybrids), 'DT-1', 'NSS', 'FEB-2' (Early blight) in pipe line (2012).

Kashi Amrit: Determinate variety derived from inter-specific cross *L. esculentum* (cv. Sel. 7) x *L. hirsutum f. glabratum* (acc. B6013') through backcross pedigree method. Fruits are round, red and weighs 108 g. TLCV resistant with average yield of 620 q/ha. Recommended for U.P., Bihar and Jharkhand.

Kashi Vishesh: Resistant to ToLCV and developed using *L. hirsutum f glabratum* B'6013' as donor parent following backcross pedigree selection method. Plants are determinate, 80 g fruitsr transplanting; yield 400-450 q/ha. Released for J&K, H.P., Uttaranchal, Punjab, U.P., Bihar, Jharkhand, Chhattisgarh, Orissa, A.P. and Karnataka, TN and Kerala.

Kashi Anupam: This is also developed by hybridization between *L. esculentum* cv. 'Sel-7' and *L. hirsutum f. glabratum* 'B6013', following backcross-pedigree selection.

Plants are determinate, fruits large, flatish round (slightly indented at blossom end of fruit), attractive red with 5-6 locules, medium maturity (75-80 days after transplanting); yield 500-600 q/ha.

Released for Rajasthan, Gujarat, Haryana.

Kashi Abhiman: Kashi Abhiman is a determinate tomato hybrid recommended at the 30th AICRP (VC) meeting held at Pantnagar, UP.

High yield potential (87 t/ha and 93 t/ha) at Ranchi and Srinagar.

Fruits are deep red in colour and mature uniformly.

The fruits are firm (pericarp thickness of 0.6 cm) and are ideal for long distance transportation.

Average fruit weight ranges from 75-95 gm and total soluble solids content range from 4.2 to 4.6 degrees Brix at red ripe stage.

This hybrid also carries Ty-2 gene that confers moderate resistance to tomato leaf curl virus disease.

Private Seed Companies Developed Several Hybrid Tomatoes

IAHS:Karnataka Hybrid, Naveen, Rashmi, Mangala, Sheetal, Rajani, Vaishali, Rupali, Ramya, etc.

Mahyco: Sadabahar, Priti, Gulmohor, Cross-B, Sonali, Morning Sun, etc.

Novartis (Sandoz) Lerica, Avinash-1, Avinash-2.

2

Brinjal

India is the primary centre of origin of brinjal. Decandolle (1904) reported it as native of Asia. In Europe it is not known earlier (before 17th early). According to N.I Vavilov, the egg plant originated in Indo–Burma region and china and spread to South Europe, china and Japan. Vavilov (1951) suggested China a centre of origin of brinjal. Crop is restricted mainly to South and South-East Asia, Southern Europe, China and Japan. Wild brinjal types are available in Southern Coastal area of Africa.

Botany

Solanum melongena' L. var. *esculentum* - round or egg shape

var. *serpentinum* – long, slender type

var. *depressum* – dwarf brinjal plant

It is often cross pollinated crop due to heteromorphic flower structure (heterostyle)

Four Types of Flowers

1. Long styled
2. Medium styled
3. Short styled
4. Pseudo styled

Only long styled and medium styled flowers set fruit.

Percentage of long and medium styled flowers is a varietal character.

Long styled –big ovary (52-90%)

Medium styled – medium size ovary (8.5-28%)

Pseudo short- styled with rudimentary ovary

True short –styled with very rudimentary ovary

Chromosome number: 2n = 24, polyploids with 2n = 36 and 48 have been produced with non significant economic use.

Genetic Diversity

Fruit shape

Long brinjal : Pusa Purple Long, Pusa Purple Cluster, Arka Shirish
Round brinjal : Pusa Purple Round
Oval : Pusa Kranthi

Fruit colour

Purple brinjal (anthocyanin) : PPL
Green brinjal (Chlorophyll) : Arka Kusmakar
Striped: Green and white : West Coast Green Round
Pink with White : Kalpatharu

Fruit shape and colour

Round purple Round white
Oval green Oval purple
Long green Long purple
Long white

Variability reported for Plant height (10-80 cm), Number of branches (3-9), First fruit set (32-91), Days to fruit maturity (56-100), Days to 50% flowering (21-50), No. of fruits per plant (9-29), No. of clusters per plant (10-29), Average fruit length (4.2-25 cm), fruit circumferances (2.7-31.2 cm) and Yield (0.27-12.57 kg/5) plant.

Germplasm

NBPGR: Activities involved in building of germplasm with wide variability.

Wild relatives: *Solanum melongena var.insanum S.torvum, S.incanum, S.indicum, S.nigram, S.khasianum, S.surratame, S.trilobatum, S.pubescence, S.khurzii, S.sismbrifolium,* etc.

Genetic Resources

1. Drought tolerance: Atabekil
2. Resistance to Phomopsis blight : Muktakesi, Bargan, White Gih
3. Resistance to Verticilium wilt : PI164941, PI174362
4. Resistance to Bacterial wilt: SM-6, SM6-6, SM-141, Cipaye, Aroman, Surya
5. Resistance to Nematode: Kalianpur T2, Long Bangalore
6. Resistance to Early blight : PPC
7. Resistance to Viral / MLO: PPR, Swati

Genetics of Characters

Character	Number of genes	Type of gene action
Fruit shape	3 genes	Round shape partially dominant to long
Prickly nature	Monogenic	Prickly dominant to non prickly i.e.: presence of spines is dominant
Fruit clusterness	Monogenic	Clusterness partially dominant to a solitary nature
Fruit size	Monogenic	Large fruit size partially dominant to small fruit size
Fruit colour	Monogenic	Purple dominant to green

Trait	Gene Dominance	References
Fruit Colour	Purple >Green	Tatebe 1944; Khan and Ramzan 1954
	Green >White	Swamy 1970; Choudhuri 1972,1977; More and Patil 1982; Gopinath *et al.*, 1986; Joshi 1989
	Monogenic	Nolla 1932; Jannaki 1972 and Ammal 1933;Swmy Rao 1970; More and Patil 1982 ; Patil and More 1983
	Two genes in complementation	Thakur *et al.*, 1969
	Complex	Fukusawa 1964; Gopinath *et al.*, 1986
	Interaction of three non-allelic genes	Fukusawa 1964; Gopinath *et al.*, 1986
Plant colour	Purple > Green	Thakur *et al.*, 1969; Patil and More 1983
Stem Colour	Purple >Green	Patil and More 1983; Wanjarl and Khapre 1977
Hypocotyls colour /seedling colour	Purple >Green mono generic	Wanjarl and Khapre 1977; Khapare *et al.* 1986
	Duplicate gene	Rangasamy and kadam 1973
Corolla colour	Purple >White or Light Purple	Wanjari and Khapre 1977;
	Interaction of three or more	Rangaswamy and Kadam1973
	non-allelic genes	Khapre *et al.*, 1986;
Leaf colour	Two complementary genes	Sharif and Habib 1977; Patil and More 1983
Leaf vein colour	Purple >Green	Khapre *et al.*, 1986;
	Three or More genes	Patil and More 1983
Flesh colour	Green flesh >white	Wanjari and Khapre 1997;
	Monogenic	Khapre *et al.*, 1986
Fruit shape	Elongated >round	Swamy Rao 1970; Choudhuri 1997;
	Round > Oval	Patil and More1983;
	Three genes control	Nimbalkar and More 1980
Bearing habit	Cluster >Single	Swamy Rao 1970

Plant spread	Erect > Spreading	Vijay Gopal and Sathumadhavan, 1973
	Two complementary genes	Khapre *et al*., 1985
Plant height	Tall> Dwarf Monogenic	Choudhuri 1972,1977
Prickles presence on petiole /Leaf / Stem	Prickles presences or absences monogeneic	Rangasamy and Kadam Bavansundarm, 1973
Male Sterility	Two recessive nuclear genes, ms1 and ms2	Chauhan, 1984
Style erectness	Incurved style and erect monogenic	Rangaswamy and Kadam,1973

Objectives: Develop varieties possessing following characters

1. High yield and Earliness
2. Erect and self staked plant habits (less branching) and plant free from lodging.
3. Good quality : Uniform fruit size, colour, thick flesh

- Fruit shape, size and colour as per consumer preference, low seed content and soft flesh

4. Diseases resistances to bacterial wilt, little leaf, phomopsis blight and root knot nematode
5. Insect resistance: Shoot and fruit borer, jassids. aphids, mites and Epilachna beetle.

   Breeding centre: PAU, HAU, GBPUA&T, KAU, IARI, IIHR, IIVR and other universities.

Breeding Methods

- Introduction
- Mass selection
- Pedigree methods
- Bulk methods
- Modified pedigree or bulk methods
- Single plant selection
- Back cross methods

Different Breeding Methods Studied for Efficiency

Progenies developed through single plant selection and mass selections were superior to those developed through pure line and single seed descent method. SSD was effective for raising level of resistance against bacterial wilt. Back cross breeding followed for transfer of resistant genes for biotic and abiotic stresses.

Pedigree selection: Following varieties developed through pedigree method of breeding

Arka Nidhi (BWR-12) : Derived from Dingrass multiple purple (DMP) x

Arka Sheel: Fruits black glossy, long and in cluster and is resistant to bacterial wilt.

Arka Keshav (BWR-21) : Derived from DMP x Arka Sheel: Fruits long purple in clusters and is resistant to bacterial wilt.

Arka Neelkanth (BWR-54): Derived from DMP X Arka Sheel. Small fruits in cluster, violet blue skin and is resistant to bacterial wilt.

Pure line selection: Following varieties developed through pure line selection

Arka Sheel : PLS from IIHR 192 from Kodagu : Oblong purple fruit

Arka Kusumakar : PLS from IIHR 193 (Karnataka) Green elongated fruits in cluster

Arka Shirish : PLS from IIHR 194-1 (Karnataka), Green extra long solitary

Heterosis Breeding

Bailey and Munson (1891) reported first time artificial hybridization in brinjal. Halsted (1901) first positive report on heterosis. Odland and Noll (1948) confirmed an increase in yield by hybrid. Days to flower and number of primary branches characters are governed by over dominant gene action. Fruit yield per plant governed by dominant gene action (Peter and Singh, 1973). Heterosis is commercially exploited for fruit yield.

Main Methods of F_1 Hybrid Seed Production are

1. Hand emasculation and pollination
2. Use of male sterility – a variety 'Blackey' reported to be male sterile
3. Induction of sterility through use of growth regulators and gamaticides : MH 100 ppm.

Purple plant: Marker available for use in production of F_1 seeds economically

F_1 hybrids developed:

Small fruited: MHB10, ABH2, Hybrid 2 and Arura.

Round fruited: Pusa Hybrid 6, NDB Hybrid 1.

Long fruited: Pusa Hybrid 5, Pant Hybrid 1, ARU 1

IIHR: Arka Navneet : F_1 hybrid IIHR22-1 X Supreme – Purple oval fruit weight 450 g.

Interspecific hybridization: Exploited for disease resitance breeding

Species	Sources of resistance
• *Solanum melongena var. insanum*	Bacterial wilt
• *S. xanthocarpum, S. nigrum*	*Phomopsis* fruit rot *S. sisymbrifolium*
• *S. torvum, S. khasianum*	Spotted beetle
• *S. indicum*	*Verticillium* wilt
• *S. sisymbrifolium*	Root knot nematode
• *S. macrocarpum*	Shoot & fruit borer
Solanum macrocarpum is tolerant to *Leucinoides arbonalis*	

S. incanum x *S. melongena* and *S. incanum* x *S. indicum* crosses were successful. One should resort to usage of bridge species and embryo rescue methods.

Interspecific Hybridization

1. *S.indicum* X *S.melongena*

   ↓

   F_1 Pollen Fertility 41.5%
   Amphidipliod it was 85%

2. *S.incanum* X *S.melongena* SM 34

   ↓

   F_1 heterosis observed for several quality traits
   Slight abnormality in metaphase -1 was observed

3. *S.incanum* (female) X *S.melongena*

   ↓

   F_1 seed with 40% germination
   Reciprocal crosses resorted with seedless fruit

4. *S.melongena* (PPL) X *S.torvum*
   ↓
   Green fruit but with reciprocal cross no success observed. Plants were resistant to bacterial wilt
5. *S.melongena X S.macrocarpum*
   ↓
   Meiotic abnormality and 16% pollen fertility was observed
6. *S.integrifolium X S.melongena*
   ↓
   Green fruit

Intergeneric grafts: Grafting was tried using genera *Solanum, Capsicum* and *Lycopersicon.* Side grafting and patch budding are observed to be successful. It can be used in growing Brinjal in problematic soils.

Other breeding methods were followed but in limited scale viz., mutation and polyploidy breeding.

Trait Profile: Selection Techniques

Selection for quality:

- Anthocyanin
- Total phenols
- Poly phenol oxidant activity
- Glycoalkolioids content

Long type varieties- rich in dry matter content, crude protein, anthocyanin, phenolic and glycoalkoloid and Oblong type-rich in TSS

Variation in varieties TSS : 1 to 3.63 °brix

Phenols: 0.044g to 0.138 g /100g (Arka Shirish)

Glycoalkoloid: 6.25 to 20.55 mg /100g (Arka Shirish)

Resistance Breeding

1. Bacterial wilt: Resistance controlled by single dominant gene (Swaminathan and Srinivasan, 1972)

   WCGR, GKVK Composite-1 and 2, Arka Keshav, Arka Nidhi and Arka Neelakanth

   Sources: Dingaras Multiple Purple, Sinampiro, Pusa Purple Cluster

   Wild species which are source of resistance: *Solanum melongena var. insanum, S. nigram,* and *S. sismbrifolium*

2. Little leaf diseases of brinjal: Crop damaging up to 40 to 80% noticed and it in controlled by MLO- Phytoplasm (spreading by leaf hoppers)

   S. viarum was Immune and *S. incanum* and *S. sismbrifolium* (8.3%) were resistant and Pusa Purple cluster was resistant

3. Phomopsis blight: *S. xanthocarpum, S. gilo, S. khasianum, S. nigram, S. sismbrifolium, S. torumand S. indicum* (highly resistant)

   Resistance was recessive and polygenic inherited (Kalda *et al.,* 1977).

4. Nematodes: Tolerant varieties – Vijaya and Black Beauty

   S. torvum and *S. seaforthianum* (R)

5. Brinjal shoot and fruit borer (BSFB):

   Free from BSFB: *S. sismbrifolium, S. integrifolium S. xanthocarpum, S. nigrum*

White Long, Throny Pendy, Black Pendy, W 165 and H-407 were highly resistant to fruit and shoot borer (Sources-Genetic and plant breeding of vegetable, K.V. Peter).

Long narrow fruited brinjal varieties were less infested than spherical fruited.

-Non preference for egg laying and larva boring easy in round fruit than in long fruits. SM 215, Banana Giant, Arka Kusumakar and SM-62

6. Jassids : Resistant / tolerant sources:

Annamalai, Pant Smarat, PPC and AM-62 and *S.khasianum* and *S.gilo*

S-188-2, PPL, S-34, S-258, Manjari Gota and Dorli.

Salient Breeding Achievements

Resistant to fruit and shoot borer: White long, Thorn Pendy, Black Pendy, W165, H407.

Resistant to bacterial wilt: SM48, SM56 and SM71, SM6-7 (Surya) West Coast Green Round, Composite 1 and Composite 2

Nematodes: Less susceptible is P8

High yielding varieties : Long fruited – PPL, H-H, PPC

Round fruited: T3 and S16

Pusa Ankur: New early variety. Semi erect, non spiny, oval round, small fruits 60-80 g, dark purple, 45 days for first picking, 33.7 t/ha.

Oval / elongated : Pusa Kranthi, Azad Kranthi

Round big fruited: Pant Rituraj and Jamuni Gola

Other Releases

1. Swarna Shree: CHES, Ranchi, PLS from local collection from Ranchi, oblong cream colour and resistant to bacterial wilt.
2. Swarna Mani
3. Swarna shyamli
4. Swarna pratibha
5. Pusa Uttam
6. Pusa Upkar
7. Pusa Bindu Brinjal

Local Varieties of Karnataka

1. West Coast Green Round: Popular in coastal districts of Karnataka, white and green stripes on the fruit, spiney, resistant to bacterial wilt. Tall plants, oval, big fruits and bears, white flowers.
2. Kengeri Local: Spineless, round to oval, light purple fruits, light pink flowers.
3. Erangeri: Green long fruits, spineless, tall erect growing plants.
4. Manjari Gota (MG): Semi-errect herb, as green coloured woody stem, green leaved. Stem, leaves and calyx are non-spiny, fruit-solitary, oblong, colour is purple with white streaks. Susceptible to bacterial wilt, the average yield - 29 q/ha.
5. Malapur Local-S: Medium, semi-erect plants, leaves and stem-green and non-spiny. Fruits solitary, occasionally very small spines, oblong, lustrous, green with broad purple stripes, purple stripe colour fades but not lusture during summer. Yield-24

Kashi Sandesh

This is a hybrid having semi-upright (height 71.0 cm) plant habit with green stems, purplish green leaves and purple flowers appear in 45 days after transplanting. Fruits are purple, size medium, shape round, fruit length 12.4 cm, diameter 10.2 cm and weight 225.3 g. The picking starts in 76 days after transplanting. Gives an average yield of 780q/ha. This has been released and notified during XII meeting of Central Sub Committee on Crop Standard Notification and Release of Varieties for Horticultural Crops for the cultivation in all the brinjal growing parts of the country.

Kashi Taru

Plants of this variety are tall and erect, height 120-130 cm, leaves and stem dark green; first flowering starts 45-50 days after transplanting. Fruits are long, purple, length 31 cm and diameter 5 cm. The picking starts 75-80 days after transplanting and gives a yield of 700-750 q/ha. This has been identified by Horticultural Seed Sub Committee and notified during XII meeting of Central Sub Committee on Crop Standard Notification and Release of Varieties for Horticultural Crops for the cultivation in U.P., and Jharkhand.

Kashi Prakash

Plants of this variety are semi upright, stems and leaves are green. Fruits are attractive with light green spots, calyx spiny, average weight 190 g. The picking starts in 80-82 after transplanting and gives an yield of 650-700 q/ha. This variety has been released and notified during XII meeting of Central Sub Committee on Crop Standard Notification and Release of varieties for Horticultural Crops for the cultivation in U.P., and Jharkhand.

Kashi Komal

This is another hybrid having semi upright plant growth habit, height 90-100 cm, with light green stem and leaves; first flowering takes in 35-40 days after transplanting. Fruits are light purple, long, soft texture, average length 13 cm and diameter 3 cm. The picking starts in 65-70 days after transplanting and gives an average yield of 800q/ha. This has been released and notified by XII meeting of Central Sub Committee on Crop Standard Notification and Release of Varieties for Horticultural Crops for all the brinjal growing parts of the country.

Brinjal : Arka Anand

- It is a high yielding F1 hybrid with resistance to Bacterial wilt.
- Suitable for Kharif and Rabi.
- Avg fruit weight is 50-55 gm.
- Yields 60-65 t/ha in 140-150 days.
- Derivative of the cross Dingrass Multiple Purple X Arka Sheel through pedigree method.
- Tall & branched plants bearing long fruits in clusters.
- Red purple glossy fruit skin with green calyx.
- Green leaves with purple leaf base and purple veins when young.
- Light purple green stem.
- Fruits tender with slow seed maturity with no bitter principles.
- Resistant to bacterial wilt.
- Duration 150 days. Yield 45 t/ha.

- A F1 hybrid between IIHR 22-1 and Supreme Green angular leaves.
- Large oval fruits with deep purple shining skin.
- Calyx green, thick and fleshy.
- Average fruit weight 450gm.
- Free from bitter principles with very good cooking qualities.
- Duration 150-160 days.
- Derivative of the cross Dingrass Multiple Purple X Arka Sheel through pedigree method.
- Tall & compact plants bearing small fruits in clusters.
- Violet blue glossy fruit skin with green purple calyx.
- Dark green leaves with purple leaf base and purple veins when young.
- Purple green stem Fruits tender with slow seed maturity with no bitter principles.
- Resistant to bacterial wilt.
- Duration 150 days. Yield 43 t/ha.
- Derivative of the cross Dingrass Multiple Purple X Arka Sheel through pedigree method.
- Tall & compact plants Bearing medium long fruits in clusters, ·
- Blue black flossy fruit skin with green purple calyx.
- Dark green leaves with purple leaf base and purple veins, when young.
- Deep purple green stem.
- Fruits tender with slow seed maturity with no bitter principles.
- Resistant to bacterial wilt.
- Duration 150 days. Yield 48.5 t/ha.
- Pure line selection from IIHR 194-1, a local collection from Karnataka.
- Tall plants, green leaves with white flowers Fruits green, extra long.
- Solitary bearing habit.
- Duration 140-150 days. Yield 39 t/ha

3

Pepper

'Chili' (European literature) word derived from 'Chile' which means any pepper in Mexico and Central America. Peppers are an important constituent of many foods adding flavor, colour, vit.C and pungency and therefore indispensable to the world food industries. China, Mexico, Japan, Turkey, Indonesia, Pakistan and India are major producers.

Origin: The cultivated chilli is of South American origin. Authorities generally agreed that capsicum originated in the new world (America) tropics and sub tropics. Pepper pods are there from 2000 year old burials in Peru. Decandolle concluded no capsicum was indigenous to the old world (Europe), peppers unknown in Europe until 16th century. Chilli was introduced into Spain by Columbus on his return trip in 1493 and cultivation spread from Mediterranean region to England by 1548 and to central Europe by close of 16th century. Portuguese carried capsicum from Brazil to India prior to 1885 and cultivation was reported in china during late 1700's.

All capsicum species have $2n = 24$. Polyploids with $2n = 36$, 48 and aneuploid $2n = 25$ reported

Karyotypes of Capsicum species: *C. annuum*, *C. frutescens* L., *C. baccatum* var. *pendulum* (WHId.) sbbaugh, *C pubescens* Kuiz & Pavon, *C. baccatum* var. *macrocarpum* (Cay.) thbaugh and *C. chacoense* ttunzlger (Olita, 1905).

Modern Taxonomy and Centers of origin: India contributed a lot to develop variability. There are 5 major cultivated species.

1. *Capsicum annuum* L:Primary centre of origin: Mexico, Secondary centre of origin: Guatemalan.
2. *Capsicum frutescens* L.: Centre of origin: Amazonia, *Capsicum annuum* and *C. frutescens* are distributed from Mexico to Central America and Caribbean region.
3. *Capsicum chinense* Jacquin: Centre of origin: Amazonia. Cultivated in South America. In U.S.A. most of the cultivars belong to *Capsicum annuum* only, except two types or varieties Green Leaf Tobasco and Tobasco of *C. frutescense.*
4. *Capsicum pendulum* Willdenow: It is grown for pickling in California. Eschbaugh reclassified it into: *Capsicum baccatum* var. *pendulum* (wild) Eschbaugh and *Capsicum baccatum* var. *baccatum* similar as *Capsicum microcarpum.*
5. *Capsicum pubescens* Ruiz & Pavon: grown in South America in a very limited scale. Both *Capsicum pendulum* and *C. pubescens* are from Peru and Bolivia.

Characters for Identification

Species	Corolla colour	Throat Spot	Corolla shape	Anther colour	Toothed calyx	Seed colour	Fl's/ node
C. annuum	White	No	Rotate	Bluish Purple	+	Tangarine	1-3
C. frutescens	Greenish White	No	Rotate	Blue	_	Tangarine	1-2
C. chinense	White-Greenish White	No	Rotate	Bluish Purple	+	Tangarine	1
C. pubescens	Purple	No	Rotate	Purple	+	Black	1
C. pendulum	White	Yellow	-	Yellow			
C. galapogense	White	No	Rotate	Yellow	-	Tangarine	1
C. chacoense	White	No	Rotate	Yellow	+	Tangarine	1
C. praetermissum	White-Lavender or Light Purple	Yellow	Rotate	Yellow	+	Tangarine	1
C. eximum	White-Lavender or Light Purple	Yellow	Rotate	Yellow	+	Tangarine	2-3
C. cardenasii	Blue	Greenish -yellow	Companulate	Pale blue	+	Tangarine	1-2

Flower structure and pollination: Flower of capsicum species are pentamerous but large fruited cultivars have 5-7 corolla lobes. Stamens alternate with the petals and correspond with them in number. **Out crossing**: less in bell pepper types and more in chilli types. According to various reports it varies from 7.62 to 36.8%.

Crossing time: any time of day light hours but best times are in early morning or in late afternoon. Anthers with two pollen sacs, pollen release is from lateral sutures longitudinally.

Time scale of female gametogenesis: Ten days time scale from meiosis to first zygotic division in *C. annuum* as reported by Dumas De Vante and Pitrat, 1977.

Meiosis	:	A-4, (-):4days before anthesis
Macrospore tetrad	:	A-3
Uninucleate embryo sac	:	A-2
2-4 nucleate embryo sac	:	A-1
8-nucleate embryo sac	:	A-0 Anthesis to pollination
Antipodal nuclei degenerates	:	A+1 pollen tubes in style
One synergid degenerates	:	A+2 pollen tubes in style
Fertilization	:	A+3
Remaining synergids degeneration	:	A+5
First zygotic division	:	A+6

Horticultural classification of pepper varieties: developed by P. G. Smith, University of California, Davis. It covers all *C. annuum* group, one cultivar from *C.f rutescens* (Tobasco) and one cultivar from *C. chinense* group particularly Rocotillo.

1. **Bell group**: Large fruited, smooth, thick fleshed, blocky or blunt, 3 to 4 lobes, square to rectangular in shape. Green at immature stage and turn red upon maturity. Some yellow or orange red at maturity. Mostly non-pungent, very few are pungent. Eg.: Non pungent (green-red): California Wonder, (Yellow-red): Golden Bell, Yolo Wonder

   (Yellowish green): Gypsy.

   Pungent (Green-red): Bullnose Hot, (Yellow-red): Rumanian Hot

2. **Pimento group**: fruits are heart shaped, large, green turning red, and smooth, thick walled, non pungent. Eg. Truhart, Pimson, Pimento L.

3. **Squash or cheese group**: fruit small to large, flat or semi pointed, smooth or rough, medium to thick walled, non pungent, medium green or yellow to mature red. Eg: Cheese, Yellow Cheese, Gambo. *C. chinense* : Rocotillo: mild pungent from Peru.

4. **Ancho group**: fruits are large heart shaped, smooth thin walled, stem indented into top and forms cup like structure, mild pungent. Eg.: Ancho, Mexican Chilli: turns green to red to reddish brown, Mulato: black green to brown black.
5. **Anaheim chilli group** (long chilli group): long fruits, slender, tapering to a point, smooth, medium thick flesh, turns green to red at maturity, moderately pungent to sweet. Eg. Moderately pungent: Sandia, Mild pungent: Anaheim chilli, Very lightly pungent: Mild California, Non pungent: Paprika (Bulgaria), In U.S.A. Paprika is a product not a cultivar. The non pungent or mild pungent Anahein types are used in the Western US to make this product and these cultivars are often called as Paprika locally.
6. **Cayanne group**: Long fruited, slender, thin walled, turns green to red on maturity, highly pungent and exceptions of non pungent varieties had been reported. Wrinkle and irregular in shape. Eg. Cayanne Long Red, Cayanne Long Slim, Cayanne Long Thick. Non pungent: Centinel and Doux Longdes Landes.
7. **Cuban group**: fruits are 3-6 X ½-2" in length and breadth, turn yellowish green to red, thin walled, irregular shape, blunt, mildly pungent. Eg. Cuban, Golden Greek, Non pungent: Italian variety: Pepperoncini.
8. **Jalapano group**: Fruit are elongated (2-3" X 1-2"), rounded, cylindrical in shape; smooth, thick walled, and turns dark green to red, highly pungent. Eg. Jalapano
9. **Small hot group**: Fruits are slender (1 ½ - 2" X ¼-1), medium to thin walled, turns green to red, highly pungent. Eg. Serrano, Fresno, Santaka.
10. **Cherry group**:fruits are small, spherical, ½-2", flattened, thick flesh, green turning to red, pungent, used for pickling purpose. Eg. Red Cherry Large, Red Cherry Small

    Non pungent: Sweet Cherry
11. **Short wax group**: 2-3" X 1-2", turns yellow to orange red up on maturity,smooth, medium to thick walled, tapering. Eg. Pungent: Floral Gem, Calaro, Non Pungent: Petite Yellow Sweet
12. **Long wax group**: 3-5" X ¾-1 ½", turns yellow to red, pointed or blunt. Eg: pungent: Hungarian Yellow Wax, Non pungent: Hungarian Sweet Wax, Sweet Banana.
13. **Tobasco group**: fruits slender, 1-2" X ¼", turn yellow or yellowish green to red, highly pungent and belong to *C.frutescens*. Eg. Tobasco, Green Leaf Tobasco.

Breeding for Horticultural Characters

1. **Earliness:** Pepper breeder Pochard (1966) observed transgressive segregation for earliness by using two methods, **A**. days taken for first blooming from transplanting and **B**. Percentage of ripe fruits harvested early. Inheritance for earliness is oligogenic (few genes).
2. **Fruit shape and size:** All the fruits with length/width ratio >1 they are classified as long. Elongated fruits are governed by poly genes which are recessive to dominant gene "O". Genes for small fruit size are dominant. About 30 genes govern the fruit size. If "O"+ 30genes are present fruit length increases and need more backcrosses (4-5) to recover the large fruit size of larger parent. Extra long fruits are associated with lower fruit yield per plant and irregular fruit poor shape and quality.
3. **Fruit quality**
   i) Taste: strong pleasing taste due to high sugar acid ratio and good flavor components.
   ii) T.S.S.: refractometer is used to measure T.S.S.
   iii) High pigmentation: polygenic dominant trait.
   iv) High vitamin 'C' content: Rymal (1983) gave quick colour test suitable for quantitative evaluation of vitamin 'C'.
4. **Flavour and pungency are independent characters:** Flavour (aroma): 2-isobutyl-3-methoxy prazine is responsible for the flavor. It is very strong and can be detected at 2 PPT in water solution or 1 drop in Olympic size swimming pool can be detected. It exists in PPB. More concentration is present in the outer wall. It is thermo sensitive. There are 23 flavor components identified in tobasco pepper other than the above.
5. **Pungency:** it is due to the presence of capsaicin $C_{18}H_{23}NO_3$, fat soluble compound, flavorless, colourless, and odourless but gives taste i.e. hotness. Capsaicin distribution in fresh Jalapano (mg/100 g).

	Fresh	Processed
Cross walls	18.37	345.96
Placenta	8.20	194.05
Seeds	0.45	128.19
Outer walls	0.12	68.24

Heating increases capsaicin by converting capsaicin precursor to capsaicin.

Quick colour test for pungency: Bit of placenta tissue is transferred to a filter paper by using a spear needle and a little of the oily secretion is absorbed by blotting paper. When a drop of 1% solution of vanadium oxy-trichloride in carbon tetra chloride (Morquis reagent) is added, a bluish colour develops if capsaicin is present. A negative colour test could be verified by taste. Permanent colour standards developed for this test. Total capsaicinoids include capsaicin and four structurally similar, pungent compounds. American spice trade association uses Scoville unit (SU) as a measure of pungency. It varies with the cultivar within the species of capsicum. Inheritance is by a single dominant gene "C". Polygenic inheritance of pungency with one major gene is also reported.

6. **Mature fruit colour:** Smith (1950)

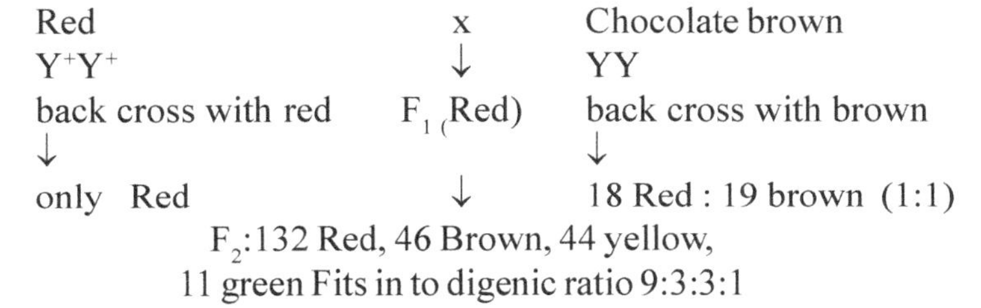

Here Y^+ represents red colour which is dominant in contrast with Y which is recessive.

Similarly Cl represents green colour which is recessive in contrast with Cl^+ which is dominant.

Y Y Cl^+ Cl^+	yellow
Y^+ Y^+ Cl^+ Cl^+	red
Y^+ Y^+ Cl Cl	brown
Y Y Cl Cl	green extends marketability in bell pepper

Immature fruit colour: The inheritance of immature fruit colour was given by Odland and Porter (1938). If all the genes are in recessive condition we get sulfury white i.e. sw_1sw_1, sw_2sw_2..............sw_nsw_n .

If one dominant gene (Sw_1^+ Sw_1^+, sw_2sw_2..............sw_nsw_n) is present we get yellowish green or lettuce green colour. Eg. Hungarian Yellow Wax. If two dominant genes (Sw_1^+ Sw_1^+, $Sw_2^+Sw_2^+$..............sw_nsw_n) are present we get dark green or cedar green colour. The dominance is incomplete so the shades of different colours are obtained. There are cultivars carrying three dominant genes. Further studies in F_2 segregation where two genes are in dominant condition the ratio obtained is 15:1 (sulfur white) and if three genes it is 63:1 by using *C. frutescens* (Odland, 1948).

Inheritance of Pedicel length: Long slender pedicel is desirable especially for bell pepper in expansion of developing fruits. It leads to deformed fruits if pedicel is short. Pedicel length is monogenic where short dominant over long.

Subramanian and Ozaki (1980) reported as long pedicel was partially dominant over short pedicel and polygenic in inheritance. At least three loci determine this character if not polygenic.

Inheritance of multiple fruitedness: *C. annuum* typically sets one fruit per axile where as *C.chinense* sets from 1 to 4 fruits/node. Transfer of this trait to *C. annuum* is associated by higher yield. Watson and Greenleaf reported that 7 semi recessive additive genes determine number of fruits/node in *C. chinense* Acc.1555.

Fruit clustering : F1 solitary, F2: 13 solitary, 3 clusters. Dominant and recessive interaction. Backcross breeding can be used to transfer these genes.

Selection for seedling emergence free from seed coat: In October-November month's lot of problems will encounter with respect to seed germination and retards growth. Promising lines were observed and this can be overcome by rigorous selection.

Internal fruit proliferation: It is a kind of abnormality in bell pepper which is associated with high temperature. A fruit like out growth from base wall/ placenta. They can be deleted by selection.

Fruiting habit : Digenic: F2 : 13 pendent 3 upright, aaB- Dominant and recessive interaction.

Plant habit: Monogenic, Tall dominant to dwarf.

Disease Resistance

1. TEV : Monogenic dominance
2. CMV : Monogenic recessive
3. TMV : Monogenic dominance
4. Nematode : Monogenic dominance
5. Damping off : Digenic

Genes

A: Dominant, Anthocyanin- purple colour foliage is incompletely dominant to green.

al-1 to al-5: Anthocyanin less- nodes are green and anthers are yellow.

B: Dominent for high content β carotene

O: oblate/round fruit shape in *C.annuum* and *C.chinense* where elongated fruits are dominant.

ms, ms-1,2,3,9,509,705: male sterile.

C: high capsaicin content-dominant

Ps (S): pod separate easily from calyx, associated with S gene to give soft flesh.

Objectives of Breeding

1. High yield and earliness.
2. Good quality – Vitamic C content high, high oleoresin content, shiny fruits, pungent and non pungent types depends on chilli or capsicum types.
3. Resitance to diseases
   a) Leaf curl complex : CMV, TMV, PVX ,PVY, etc.
   b) Anthracnose : *Colletotrichum capsaicii*
   c) Powdery mildew
   d) Phytophthora rot.
4. Resistance to insects
   a) Aphids
   b) Thrips
   c) Borer

Methods of Breeding

1. Introduction

Specially in capsicum : California Wonder, Yolo Wonder ,etc.

2. Pure line selection

Arka Lohit developed through Pure Line Selection from IHR 324 a local collection. Arka Abhir developed through Pure Line Selection from Dyavanur dabba. Pure line selection has been used to develop varieties like G1, G2, G3, G4, NP-46-A, K-1, CO-1, Mussalwadi, Shankeshwar-32, Makapuri, Sindhur, CA-1068, Seema Mirapa, Patna Red, NP-34, NP-41, etc.

3. Mass selection

Arka Basant-Mass selection from Var Soroksari, Arka Mohini-Mass selection from Titan. Arka Gaurava-Mass selection from IHR 214-1 (Golden Calwonder).

4. Pedigree method

The varieties developed are G5, K-2, Punjab Lal, X-235, X-206, Pusa Jwala, Pant C-1, Pant C-2, Jawahar Mirch 218.

5. Mutation breeding

The mutagens used are EMS, DMSO, gamma rays, X-rays. Eg.: MDU-1 has been developed at Agriculture college and research institute, Madhurai by treating the seeds of K-1 with gamma rays. Released by TNAU during 1977. "Albena" chilli variety developed through mutation. Other mutants developed by gamma rays (12 kr) and N-Nitroso-N-ethyl urea treatments are effective. It is to create variability and scope is little as it is not directed method..

6. Heterosis breeding

High hetorisis was observed between different genetic groups. Grouping was done based on D^2 values clustered into 9 gene constellations. Canonical variate analysis and metroglyphs confirmed grouping. Heterosis was observed for earliness, plant height, fruit girth, number of fruits and yield. CH1 hybrid from PAU, Ludhiana which is (genic male striking) first chilli hybrid from Government organization. Used gms (genic male sterile) line as female for F1 seed production. Many private hybrids in chilli and capsicum.

PAU: Punjab Hybrid-1 (CH-1): MS-12 X LLS (released during 1992), CH-3.

IIHR, Bangalore: Arka Meghana (+PM, viruses), Arka Harita (+PM), Arka Swetha (+PM); all the three varieties were developed by using CGMS (Cytoplasmic genic male sterile) lines.

F1 seed production systems

1. Hand emasculation and pollination : Capsicum and capsicum x chilli for chilli seed production.
2. Protogyny / protandry : It can be utilized with markers.
3. Genic male sterility: Hybrids from PAU, Ludhidana, CH1 & CH3 this system is used.
4. Genetic cytoplasmic male sterility : Hybrids from IIHR, Bangaluru this system is used. Eg. Arka Meghana, Arka Harita
5. Selective differential gameticide : Spraying of FW-450 at 0.3% concentration for 1 to 5 times can selectively kill male gametes. This needs to be used with markers as fertile flowers can exist and contaminate because of continuous flowering and limited period effect of these gameticides.

7. Interspecific hybridization

1. *C. frutescens:* Source of resistance to *Phytophthora capsici*, Cercospora leaf spot, leaf curl, mosaic and bacterial wilt
2. *C. baccatum* var. *Pendulum:* Source for Powdery mildew
3. *C. annum* x *C. chinense :* Progenies had high value of capsaicin (0.92%), Oleoresin (34.4%) and colour (58.62 ASTA)
4. *C. annuum* x *C. baccatum*-Powdery mildew resistance transferred to *C. annuum* with indeterminate growth habit under both chilli and capsicum background at IIHR, Bangalore.

8. Polyploid breeding

Natural polypoids 2n = 36, 48 are available in bell pepper group. Which have low fertility rate and seed content is very low. Polyploids synthesized by using colchicine. They have limited value.

Other methods like Bulk population method, Backcross breeding, etc are being followed and also suggested to follow depending the objectives and genetics of the trait sought to be improved.

Bell Pepper (Sweet pepper)

It is the horticulture type of *Capsicum annuum* L and has high percentage of self pollination compared to chilli (Hot pepper) and genetics of various traits are as applicable to chilli and these two corps form just two populations of same species. Chilli traits (pungency and fruit size) are dominant over sweet pepper and these two can be crossed easily and hence while providing isolation distance at the time of seed production these two crops should be considered as one crop.

Floral biology: Anthesis starts at 7:15AM & continues up to 11:15AM, with peak at 7:15AM. Anther dehisces after 30 min of anthesis. Stigma becomes receptive from the day of anthesis & remains receptive upto 2 days after anthesis.

Genetics: Fruit size is controlled by additive genes. Number of days to flowering is controlled by dominant genes. Plant height, no. of fruits per plant and total yield by over dominant genes. Days to flowering, days to first harvest and early yield per plant is controlled by additive and non additive effects, duplicate epistasis was also present. Pungency is controlled by dominant genes. Calyx enclosing fruit base is controlled by single dominant gene. Selection for earliness is based on days to flowering and days to fruit harvest.

Breeding Methods in Bell pepper

Introduction: direct introduction which have been promising are Yolo Wonder (Tolerant to TMV), California Wonder, Chinese Giant, R-449 (Russian Introduction) and World Beater performed well.

Pure line selection: Arka Mohini from Titan (USA), Arka Gaurav from Golden Calwonder (USA), Arka Basant from Soroksari (Hungary).

Pedigree selection: Being followed to develop varieties

Back cross method: Being followed for transfer of one or few genes. Eg. Resistance to powdery mildew from *C. baccatum* var. *pendulum*.

Heterosis breeding: F_1 hybrids are popular in USA, Europe and other countries. Bharat F_1 hybrid from IAHS has met with success with growers (absence of male sterility). Heterosis over standard varieties has been reported. Heterosis in capsicum is at lower magnitude compared to chilli because of narrow genetic base. Male sterility reported is unstable and need to use seedling markers. Eg. IARI, Katrain: Pusa Deepti.

Mutation breeding: Several mutants including male sterile mutants have been identified.

Breeding for Disease Resistance

1. **Anthracnose / fruit rot / dieback**: *Colletotrichum* sps Yield decreases drastically and reduction can be to the extent of 65 per cent. It is prominent in chilli over capsicum in all tropical and subtropical chilli growing areas of India. Sources: Lorai, Bangla Green, Perennial, S-27.
2. **Viral diseases**: Most common in all tropical and subtropical chilli growing areas of country. Yield loss 80-100%: TMV (several strains), CMV, PVY, PVX, Chilli mosaic virus, leaf curl virus, SLTMV (Samsun Latent strain of TMV). Resistant sources for TMV, CMV, LCV: Pant C-1, Punjab Lal, Laichi, Tawari, Bangla Green, Lorai, Perennial, Longi. IARI: Delhi Local, 38-2-1, 96-4-9-3, S-20-1, 141-2-33.

   Chilli leaf curl is a complex disease where, TLCV (tobacco leaf curl virus), PMV (pepper mottle virus), mites, thrips and aphids are possible causes and resistant varieties are Myliddy1 and Myliddy2.

   Mosaic complex where PV-Y, PV-X, TMV, Tabaco etch virus and CMV are posible causes and resistant varieties are A-127, C1 and Puri Red.
3. **Powdery mildew**: Caused by Levellula taurice where reported 34% yield losses. Serious in southern parts of India. Sources: EI Salvador, *C. frutescens*, *Capsicum baccatum* var. *Pendulum*
4. **Bacterial leaf spot**: *Xanthomonas campestris* fsp. *Vesicatoria* is the causal organism and is serious in southern parts of India. Symptoms are defoliation, scorching of leaves. Sources of resistance are *C. annuum*, PI 163189, PI 322719 and PI163192 and are resistant to race 1 and race 2. PI 322719 is more valuable with larger and non pungent fruits.

   PI 271322 X Early Calowonder

   ↓

   $F_1 \rightarrow F_2$

   140 susceptible and 10 resistant plants.

   Two recessive genes are involved for resistance.
5. **Cercospora leaf spot:** Sources of tolerance are IIHR-328, 344-9.
6. **Phytophthora fruit rot**: seedlings can be inoculated with the inoculum and planted in the main field for screening. Yolo Wonder α Carries dominant

resistant genes. F_1s with susceptible check shown resistance, F_2 segregation of 3 : 1 and 15 : 1 ratios so one or two genes involved which are duplicate. Seedling root inoculation tried as artificial screening method. Pochard concluded that PI 201234 is the best source of resistance and is determined by at least two complementary dominant genes. Mulato is a highly resistant source from Bulgaria.

7. **Nematodes**: RKN- Santaka and 405 B Mexica (*C. annuum)* each carry a single dominant gene "N" and is resistant to Southern Root Knot nematode race 'Acrita' identified in *M. incognata*.

Multiple disease resistant sources: MS-12, 13, 41, 11-1, 41-1, Tiwari, Laichi.

TMV: -L- Local necrotic lesions, l: recessive

l^i- scattered secondary lesions (dominant for European strains)-recessive to L.

Mechanically transmitted to capsicum from other crops like,

Tomato: CMV, TEV, TMV, PVX.

Cantaloupe: CMV.

Egg plant: CMV, TEV.

Tobacco: TMV, TEV, PVX.

Mustard: PVX.

Aphid (*Myzus persicae* & *Aphis gossypii*) transmitted viruses: CMV, PVY, TEV, PMV.

TSWV is transmitted by thrips.

Seed transmitted viruses: TMV, SLTMV.

Pepper breeding programme: Backcross- intercross for CMV & TMV resistance in bell pepper

XVR-3-25 : et^a- resistant to TEV;

L^1 : imperfect localization of TMV;

Bs1 : resistant to bacterial spot Race1;

Bs2 : resistant to bacterial spot Race2.

C. chinense: **Acc.1555**:- immune to PMV and carried resistant gene et^{c1} (resistant to PMV & TEV-C) or et^{c2} (resistant to PVY, PMV & TEV-S. Additional desirable trait of Acc.1555 are round fruit shape (gene O), earliness, ripe fruit rot resistant, multiple fruitedness. **XVR-3-25:** Large fruited and multiple disease resistant and is also highly resistant to CMV. **Perennia**l: source of resistance to CMV.

Ra- escapes infection at low inoculums dose;

Rb- hypersensitive type of resistance;

Rc- non necrotic and slow rate of virus multiplication.

Backcross-Intercross scheme for breeding PMV- and CMV reststant bell peppers.

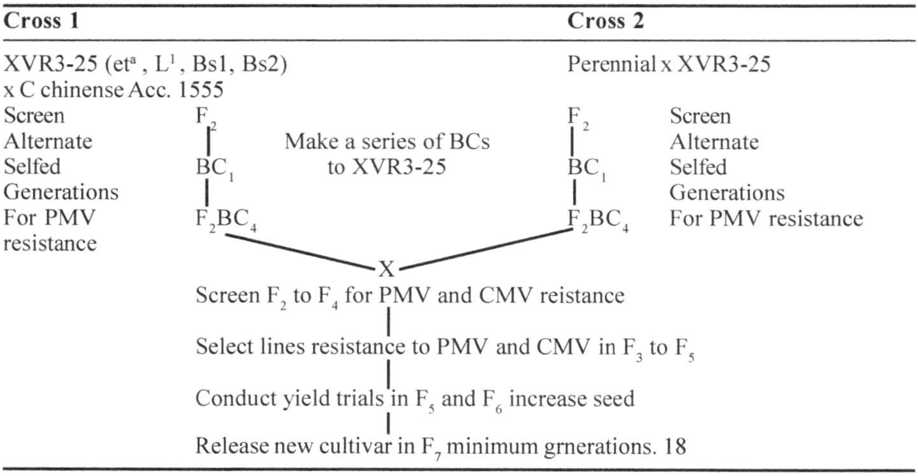

Cross 1			Cross 2	
XVR3-25 (et[a], L[1], Bs1, Bs2) x C chinense Acc. 1555			Perennial x XVR3-25	
Screen Alternate Selfed Generations For PMV resistance	F_2 → BC_1 → F_2BC_4	Make a series of BCs to XVR3-25	F_2 → BC_1 → F_2BC_4	Screen Alternate Selfed Generations For PMV resistance

X

Screen F_2 to F_4 for PMV and CMV reistance

Select lines resistance to PMV and CMV in F_3 to F_5

Conduct yield trials in F_5 and F_6 increase seed

Release new cultivar in F_7 minimum grnerations. 18

Virus Screening Techniques

Carborundum leaf wiping method: reliable for even less infectious viruses. Inoculum is prepared by grinding leaves or whole shoots of infected source plants in a mortar or blender. A weighed sample is ground in a measured volume of water or phosphate buffer (KH2PO4 0.01 to 0.1 M at 7.0-7.5 pH). Crude extract is filtered through cheese cloth. TMV extracts are used at 1/100 and TEV, PVY, PMV and CMV at 1/10 to 1/20 W/V dilution. The leaves to be inoculated are marked and dusted with 400-600 mesh carborundum. Avoid contamination. Wash hands with soap and water. Spray gun method of inoculation: this method works well with highly infectious viruses like TMV or SLTMV, but is not as reliable with the less infectious viruses like TEV, PVY, PMV & CMV. Carborundum must be added to the inoculums about 5% by volume. Plants are spayed forcefully from a distance of 3 to 4 inch at 60-100 psi. Generally 5 to 6 weeks old seedlings can be inoculated in field. ELISA: Enzyme Linked Imminosorbent Assay is run to identify the virus and strain.

Multiple Pest and Disease Resistant Varieties

Punjab Lal: +CMV, CMMV, wilt & Spodoptera leaf eating caterpillar.

Pant C-1: mosaic & leaf curl virus.

Pusa Sadabahar: +CMV, TMV & leaf curl.

Breeding for Insect Resistance

Lines tolerant to thrips: Caleapin Red, Chamatkar, NP-46-A, X-1068, X-743, X-1047, BG-4, X-226, X-233, X-230.

Lines tolerant to mites: LEC-1, Kalyanpur Red, X-1068, X-204, Goli Kalyanpur-309-1-1-15, 300-1-5-1, S-118 (Punjab Lal), 635, 565.

Lines tolerant to aphids: LEC-28, LEC-30, LEC-34, Kalyanpur Red, X-1068.

Resistant Chilli varieties

Fruit rot (Colletotrichum capsici) Leaf curl virus	K-2, Pusa Jwala, Pusa Sadabahar, Pant C-I
Leaf curl, CMV, TMV, Wilt & dieback	Punjab Lal, Punjab Surkh
Bacterial wilt	Utkal Rashmi, AAUM-1 , AAUM-2
Powdery mildew	Arka Suphal, Arka Harita, Arka Meghana
Aphids	Kalyanpur Red, X-1068
Mite	Punjab Lal, Pusa Jwala
Thrips	Chamatkar, NP-46A, X-1068, Pusa Jwala

Disease	Resistant variety
Damping off	Kuban early
Fusarium wilt	College No.9
Powdery mildew	Avelar
Bacterial leaf spot	Santaka
Bacterial wilt	Kandari
Leaf curl (Tobacco leaf curl virus)	Pant C1, Puri Red, Puri Orange
TMV	Florid Vr2, Rutgers
PVX	Florid Vr2, Rutgers
Tobacco etch virus	Florid Vr2, Rutgers
CMV	Lr1 and X-196
Nematode	Bombay 742, Santaka XS

Breeding Achievements

Chilli varieties / hybrids

Chilli F1 hybrids from IIHR

Arka Sweta (MSH 149): A male sterile (CGMS) based chilli F1 hybrid. Fruits dark green (38 t/ha) turn to deep red and yields 6 tones dry chilli per hectare.

Arka Harita: Chilli F1 hybrid developed by using male sterile (CGMS) lines. It is resistant to powdery mildew and viruses.

Arka Meghana: Chilli F hybrid developed by using male sterile (CGMS) lines and it is tolerant to powdery mildew and viruses.

Arka Suphala : Resistant to Powdery mildew

Other varieties

1. Puri Red x Local
   ↓
   NP46A
   Selected in advanced generation
2. NP46A x Local Kandari
   ↓
   Pant C1 (Natural cross): Resistant to leaf curl and Bacterial wilt 4 isolates.
   Selected in advanced generation
3. NP46A X Puri Red
   ↓
   Pusa Jwala: Selected in advanced generation

Arka Lohit :(IIHR, Bangalore): A pure line selection from IHR 324(local collection). Plants tall. Fruits dark green, smooth, straight, turning deep red on maturity (Capsanthin 0.205%). Fruits highly pungent (Capsaicin 0.708%) Suitable for irrigated and rainfed cultivation. Duration 180 days. Yield 25 t/ha. (green Chilli). 3 t/ha. (dry chilli).

K-1: It is a selection from an Assam type B72A. Tall and compact, fruits are 6.6 cm long and shiny red with capsaicin content of 0.35 mg/g. Yields 1900 kg dry pod/ha in 215 days.

K-2 (Kavilpatti, TN, 1985): Hybrid derivative of B70A and Sathur Samba. Plant height 90 to 95 cm, Tall and bushy, fruit length 6.1 cm and 4 cm girth, bright red, tolerent to thrips, yield 0.75 t/ha, capasaicin 0.49 mg/g.

Bhagyalakshmi (Lam, AP, 1977): **G-4** is a Samba type known as Bagyalakshmi in Andhra Pradesh. Average plant height 52.9 cm. Leaf lamina dark green. Fruit 8.2 cm long with 0.7 cm width. Crop duration is of 180 days. Yields 1.6 t/ha dry chillies under rainfed and 4.5 t/ha under irrigated conditions. Plants are tall, dense and fruits are 8.8 cm long. The pedicel is attached with fruit firmly. Fruits are bright red, 0.52 mg capsaicin per g of fruit.

Pusa Jwala (IARI, 1977): Developed from a cross of NP-46A and Puri Red. Plant dwarf and bushy with light green and broad leaves. Fruits long (9 to 10 cm) with light green colour. Highly wrinkled, pericarp thin and light red in colour. High yielding, resistant to TMV. Average dry yield is 2 to 2.5 t/ha. Capsaicin content is 0.48 mg/g of fruit.

Sindhur or CA-960 (Lam, AP, 1978): Fruits long (9.2 cm) with broad base with semi-shaped calyx. Pericarp thick with light green colour, turning to attractive red colour, pungency mild, suitable for green and red chilli production. Average yield 1.2 t/ha (rainfed) & 6 t/ha (irrigated) in 180 days. Dual purpose.

X-235 (Lam, 1985): Early maturing plant with short internodes. Leaves small and dark green. Flowers with yellow anther, fruit 5 to 6 cm long, calyx deeply cup shaped. Fruit tip pointed, highly pungent, seed percentage 45%. Maximum yield potential up to 7.5 t/ha dry chilli.

Pant C-1 (Pantnagar): Advanced selection from cross NP46A and Local cultivar resistant to leaf curl and mosaic virus. Plants dwarf (50-60 cm) and bushy, erect. Fruit medium size (6 to 7 cm), light green at immature stage and red at maturing. Fruits smooth with blunt apex. Resistant to leaf curl virus. Average yield is 2 to 2.2 t/ha.

Punjab Lal (PAU, Ludhiana, 1985): Plants dwarf and bushy with dark green leaves, fruits erect, medium sized, dark green turning dark red glossy at maturity. Average dry yield is 2 to 2.2 t/ha. Rich in capsaicin, oleoresin, dry matter and pigment and highly suitable for dry powder production. Resistant to TMV, CMV, PVY and leaf curl viruses. Moderately resistant to fruit rot and die back diseases.

CO-1: Selection from a 'Samba' type from Sathur of Ramanathapuram district in Tamil Nadu. Plants erect, medium tall and compact with moderate branching habit. The fruits measure 7.3 cm long with a capsaicin content of 0.72 mg/g, yields 2.1 t/ha in 210 days.

CO-2: A gundu type selection from CA (P) 63 is suited for both green and dry chilli. The plants are medium tall and erect and the fruits are thick red with more seeds. Suited for all three seasons. Yields 4.35 t/ha green chilli and 0.87 t/ha dry chilli.

MDU-1: An induced mutant obtained by gamma irradiation of K-1 cultivar. Plants tall and compact with large broad leaves and with clustured flowering habit (4 to 9 flowers/node). Fruits are long, shiny with a capsaicin content of 0.7 mg/g. It yields 1809 kg dry chilli per ha in 210 days.

Pant C-2: Selection from cross of Pant C-1 Parents (NP46A and Local). Plants are tolerant to mosaic and leaf curl virus, fruit pendent and yields 1.47/ha (dry).

JCA-154 (J.N. Krishi Viswavidyalay, Jabalpur, MP): Pickling cultivar, fruits are dark green when unripe and bright red when ripe. Fruits mature in 115 to 120 days.

Punjab hybrid-1: F1 hybrid where male sterility (MS-12) is exploited for seed production.

NP46A (IARI): Plant dwarf and bushy with light green leaves. Fruits are long (8-9 cm) and thin, wrinkled with high green colour, turning light red on ripening. High yielder, tolerant to thrips. Cultivated widely. Average dry yield is 1.5 to 2 t/ha. Capsaicin content is 0.53 mg/g of fruit.

Andhra Jyoti: (Lam. AP, 1977): **G-5:** This is evolved from a cross between G-2 and 1331, which is tolerant to thrips. The fruits of this cultivar are short and gundu type. Plants are tall and dense and the fruits are bright red measuring 5.1 cm in length and 6.3 cm in girth. The capsaicin content is 0.65 mg/g of fruit. Fruit medium pungent with thick glossy red pericarp. Average dry yield is 1.14 t/ha under rainfed and 5 t/ha under irrigated condition.

Pusa Sadabahar: Cultivar suited for green, red chilli and preparation of oleoresin. It bears fruits in clustar of 6 to 14. Green chilli yield is 7.5 to 10.0 t/ha and dry yield 1.5 to 2.0 t/ha.

Ujwala: Bacterial wilt resistant variety developed at KAU, clustered, erect and dark green fruits turning deep red on ripening, 9 to 10 fruits/cluster. Average fruit weight is 2.5 g, length 6.2 cm and girth 3.8 cm and 357 fruits/plant. Fruits contain Oleoresin 24 per cent, capsaicin 0.49 per cent with a colour value of 139.29 AST units. It yields 26 t/ha (green). First picking starts after 90 to 110 days.

Jawahar 218 (Jabalpur, 1987): Plants dwarf and spreading, light green leaves, fruits are long (10 to 12 cm), wrinkled, light green at immature stage and rosy red at maturity. Early yield 1.8 to 2.2 t/ha. Suitable for green and red chilli production. Tolerant to leaf curl and fruit rot.

Jawhar mirch 283 A summer set chilli variety released from JNKVV, Jabalpur Besides, local types like Byadagi kaddi, Sankeshwar, Savanur Dabbi, Chincholi Local, etc., from Karnataka, and private varieties like Tejaswini (Mahyco), Delhi hot, etc., are there.

Ujwala (CA-219): High Yield (18t/ha), Clustered, linear and erect fruits, Average fruit weight : 2.54 g, High pungency

Jwalamukhi (Cul. 57): High Yield (18t/ha), Pendulous, medium long, thick skinned fruits Average fruit weight : 6.5 g, Low pungency.

Jwalasakhi (Cul. 45): High Yield (19.6 t/ha). Pendulous, medium long, thick skinned, smooth tapering fruits. Average fruit weight : 5.2 g. Low pungency

Arka Abhir : A pure line selection from Dyavanur dubba in Kundgol, \Dharwad District). Plants tall. Fruits light green, wrinkled turning deep red on maturity. Has better colour (color value maximum 1,65,541 c.u) and low pungency(0.05% Capsaicin) . Suitable for oleoresin extraction, oleroesin yield from fruits without seed 5.78%. Duration 160-180 days. Yield 2.0 t/ha (dry Chilli).

Arka Suphal (PMR 57) : Indeterminate with a plant height of 80-90 cm. Dark green foliage, fruits straight, smooth with pointed tip, 7-9 cm long. Fruit colour green changing to deep red Yield: 25t green and 3 t dry chilli/ ha. It is resistant to powdery mildew and field tolerant to viruses.

MSH 206 has been developed by the institute. This is a CMS based F1 hybrid, fruits smooth and acceptable quality with medium pungency, yields 40-45 tones /ha fresh green or 5-5.5 t/ha red dry fruits in 180 days. The hybrid is moderately tolerant to viruses and suitable for fresh markets.

Arka Sweta: High yielding F1 hybrid developed by using MS line. Plants medium tall (95 cm) & spreading (82.5 cm). Fruits long (13.2 cm) with 1.3 cm width. Fresh yield 38.4t/ ha and dry yield of 6 t/ ha in 140-150 days. Fruits are light green, turns red.

Arka Meghana: High yielding chilli F1 hybrid developed by using MS line. Plants medium tall (81.3 cm) & spreading 69.5 cm. Fruits long (10.6 cm) with width of 1.2 cm. Very early, taking 24 days for 50% flowering. Fresh yield of 33.5 t/ ha and dry yield of 5 t/ ha in 140-150 days. Fruits are dark green and turn deep red. Tolerant to powdery mildew and viruses.

Arka Harita : High yielding chilli F1 hybrid developed by using MS line. Plants tall (1m) & spreading (90cm). Fruits medium long (10 cm) with width 1 cm. Fresh yield 31 t/ hectare and dry yield 6 t/ ha in 150-160 days. Fruits are dark green and turn red. Tolerant to powdery mildew and viruses.

Capsicum varieties/hybrids

Pusa Deepti, California Wonder, Chinese Giant, World Beater, Yolo Wonder, Arka Basanth, Arka Gaurav, Arka Mohini, Bharat (IAHAS) are some varieties /hybrids of hot pepper (capsicum) developed and released for cultivation .

California Wonder: Introduction from the USA at IARI Regional Station, Katrain. Area of adaptation. All over India Plant habit. Vigorous, upright prolific bearer. Days to first picking 90-100 days after transplanting . Fruit characteristics : Smooth, fine flavoured with 3-4 distinct lobes, mooth, thick flesh, deep green, turns bright crimson at maturity. Yield :170 q/ha

Yolo Wonder : Plant habit: Late, plant upright dwarf and bushy. Days to first picking: 100-110 days after transplanting , Fruit characteristics: Smooth, fine flavoured with 3 fused lobes, smooth, thick flesh, deep green, turns bright Yield: 160 q/ha .

Arka Basant: Pedigree: Improvement over IIHR 225-1-1 (var. Soroksari) followed by mass selection (Sel 15). Year of release: 1984 by SVRV and 1986 AICRP. Area of adaptation : All over India . Plant habit: Indeterminate. Fruit colour : Cream colour which turns orange red on ripening. Fruit shape: Erect conical, thick fleshed, 2-3 lobed. Crop duration: 130-160 days. Fruit weight: 50-

80g. Yield : 150-200q/ha. The variety has very good cooking quality and export potential.

Arka Mohini (Sei-13) : Pedigree: Improved over IIHR 3 12-1-2 (Titan variety) followed by mass selection. Year of release : 1984 by SVRC and 1986 by AICRP. Area of adaptation: All over India. Growth habit: Determinate. Crop duration : 135 days. Fruit characteristics : Thick fleshed, 3-4 lobed dark green blocky fruits, average fruit weight (180-200 g), fruit are pendent , which turn red on ripening, Fruit yield: 200 q/ha, Seed yield : 150 kg/ha .

Arka Gaurav (SeI-16): Pedigree: Improvement over LIHR 214-1(Golden Calwonder) followed by selection. Year of release: 1984 by SVRC and 1986 by AICRP. Area of adaptation: All over India. Growth habit : Indeterminate, Duration of the crop: 140 days, Fruit colour: Dark green blocky fruits which turn orange yellow on ripening, Fruit size : Medium, 3-4 lobed, Fruit weight : 70-80 g Yield : 160-200q/ha, Tolerant to bacterial wilt

---

Hybrid-Kashi Early	This F1 hybrid has been developed by crossing PBC- 473 x KA-2 at IIVR Varanasi. Plants of are tall (100-110 cm height) without nodal pigmentation on dull green stems and bear pendant fruits. Fruits are long (8-9 x 1.0-1.2 cm), attractive, dark green and turn bright red at physiological maturity, pungent with smooth surface. First picking of the green fruits starts at about 45 days after transplanting. The average yield of this hybrid is 250 q/ha (red ripe). It has been recommended for growing in states of Uttar Pradesh, Bihar, Madhya Pradesh, Jharkhand, Uttaranchal, Karnataka, Delhi, Punjab, Haryana, Andhra Pradesh and Chattisgarh.
Kashi Anmol	This is an improved population derived from two cycles of simple recurrent selection from a Sri Lankan introduction. Plants are determinate, dwarf (60-70 cm) with nodal pigmentation on stem and bear green attractive pendant fruits. First picking starts 55 days after transplanting; yield 200 q/ha. At the farmers field has given an average green fruit yield of 250 q/ha in only 120 days of crop duration. This has been identified for release and cultivation through AICRP for green fruit production for the cultivation in Punjab, U.P., and Jharkhand.
Kashi Surkh	This is a F1 hybrid of a cross between cms line (CCA 4261) and inbred derived from Pusa Jwala. Plants are semi determinate (1-1.2 m), erect and nodal pigmentation on stem. Fruits are light green, straight, length 11-12 cm, suitable for green as well as red fruit production. First harvest starts after 55 days of transplanting. Green fruit yield is 240 q/ha whereas red fruit yield is about 140 q/ha. This is tolerant to thrips, mites and viruses. This has been identified by AICRP for the cultivation in West Bengal, Assam, Punjab, U.P., Bihar, Jharkhand, Chhattisgarh, Orissa, A.P. Rajasthan, Gujarat and Haryana.

Punjab -27: Developed at PAU, Ludhiana, Year of release: 1974, Fruit colour : Light yellow Fruit shape : Elongated, Taste Sweet, Yield : 75q/ acre, Miscellaneous : Set seeds in plains of Punjab.

Nishat -1 : Developed at SKUAT, Srinagar, Area of adaptation : Recommended for cultivation in zone I, Growth habit: Indeterminate, Days of first harvest: 45-50 DAT. Fruit colour : Dark green black, turn yellow at physiological maturity, No. of lobes: 34, Yield: 200-300 q/ha, has long shelf life.

Pusa Deepti (F1 Hybrid) : Developed at IARI Regional Station, Katrain, Year of release: 1997, Area of adaptation : Himachal Pradesh and J & K, Plant habit : Erect, medium and bushy, Days to first picking: Early, 60-70 days after transplanting. Fruit characteristics :Yellowish green, conical, turns dark red at maturity, 9-11 cm long. Yield : 225 q/ha

Kt-Pl-19: Paparika type variety developed at lARl Regional. Station, Katrain, Pedigree: Pure line, selection from P-12, Year of release: 1994, Area of adaptation: Karnataka, Himachal Pradesh and colder regions of Tamil Nadu. Plant habit: Plants upright, Fruits shape: Conical, flat, Fruit length: 13-17 cm long, Fruit diameter : 2.2-3 cm, Fruit colour: Dark green, pendent, turn dark red at maturity. Suitable for powder and having good export quality.

Lario (F1 hybrid): Source: Developed at Syngenta India Ltd., Pune, Year of release: Recommended for cultivation in zone lV-VIII, Plant characteristics: Plants are medium tall, compact vigorous, dense foliage provides adequate shelter to fruits, Days to first harvest: 60-65 DAT, Fruit colour: Dark green, Fruit size: Medium, Fruit weight: 150g, No. of lobes : 34, Yield: 300-450 q/ha, Requires stacking.

Aporva (CP-40): Mild pungent capsium variety is an improvement over local pungent types from Northern Karnataka, developed at University of Horticultural Sciences, Bagalkot by Dr Ravindra Mulge. Fruits are light green with more number of ribs and bears more freuits with an yielding ability of 10 to 11 t/ha.

4

Potato

If there is scarcity of food, potato can come to the rescue because of its virtue of high yielding (40-50 t/ha) and is a staple food. It produces more calories per unit area and per unit time than any other major food crops. Potato which is the staple food of Europeans and Americans belong to the highlands of Andean mountainous regions of Peru and Bolivia where it has been cultivated since 6000 years.

Origin: Peru, Bolivia. Many wild species occur at Mexico and Central America. Potato was introduced to Europe in 16th century by Spaniards (Salaman, 1966) and in early 17th century to India most probably by Portuguese. There are some suggestions based on linguistic and botanical evidences that potato might have been introduced in India directly from South America across the Pacific Ocean. Earlier potatoes were grown in hilly areas of North and South since British settlers brought the varieties continuously from UK and these were bred for long day temperate conditions of Europe. But in India potatoes are grown in short day-subtropical plains. The seed stocks used to degenerate due to accumulation of viruses and loose their yield potentiality. Seed potatoes were mainly imported from Europe i.e. U.K., Italy and Netherlands and Burma till the beginning of Second World War and the imports gradually reduced.

Classification and origin of the cultivated potatoes

Species	Distribution	Origin
Solanum tuberosum sub sp. *Tuberosum* (2n = 48)	Cosmopolitan	By artificial selection in Europe, North America and Chile from introduced clones of group 2.
S. tuberosum sub sp. *andigena* (2n = 48)	Venezuela to northern Argentina; also sporadically in central	America and Mexico From group 4 and 5 by spontaneous doubling of the chromosome number.
S. chaucha (2n = 36) S. *phuerja* (2n = 24)	Central peru to northern Bolvia Venezuela to northern Bolvia	By hybridization between group 2 and 4 and 5. by selection for short tuber dormancy from group 5.

S. stentomum (2n=24)	Southern peru to northern Bolvia	By natural hybridization between wild species followed by artificial selection.
S. juzepcsukii (2n=36)	Central peru to southern Bolvia	From crosses of S.acaule with groups of 4 and 5
S. curtilobum	Central peru to southern Bolvia	Crosses of S. juzepczukii with group 2.

Potato Breeding History in India: Organized research on potato started on 1st April 1935 in Imperial Agricultural Research Institute, New Delhi and opened the research station for potato breeding in Simla, but got fillip when CPRI, Patna was started by MOA, India in 1949 with many objectives mainly breeding and selection of High yielding disease resistant cultivars and adaptability to different agro-climatic regions of India. During 1956-CPRI was shifted to Shimla where flowering is more due to long photoperiods and low incidence of viruses. During 1968-CPRI control was transferred to ICAR. During 1971-ICAR setup All India Coordinated Potato Improvement Programme (AICPIP) with its head quarters in Shimla. This project is funded by ICAR majorly and SAU's. Major development in potato is development of seed plot technique by Pushkarnath (1967) to multiply the seed or breeding material under disease free conditions. By 1984 total of 26 high yielding varieties were developed for different agro-climatic regions of India (Mishra *et al.*, 1984).

About 80% of potato is grown in short day conditions in India from October-March. Therefore, the major thrust is to develop the varieties suitable to different regions and day length conditions. Notable scientists worked on potato in India i.e Pal, Ramanujan, Mehta, Pushkarnath. Hybridization programme started in CPRI with three regional stations i.e Jalandhar in Punjab, Meerut in Patna and Modipuram. Future thrust is on Biotechnology for developing varieties that are salt tolerant, fertilizer efficient and nutritionally effective.

Botany: Basic chromosome number x = 12 and 2n = 24, 36, 48 60 and 72 observed. But 2n=48 is commonly cultivated.

It flowers under long photoperiod in hills. Efforts are made to induce flowering under day neutral and short day conditions. True Potato Seeds (TPS) where fresh seeds have dormancy for 5 to 6 months. Break dormancy by seed soaking in 1500-2000 ppm GA3 for 48 hours.

Areal stem behaves as an annual but plant is perennial because of its underground stem called stolon. Stolon has nodes and internodes under favourable conditions develop into branches which grow more or less horizontally outwards.

Reproductive Biology

Flowers under long day, moderate temperature and high relative humidity (short day for tubrization). Long days are available in hills of North India where it is grown under long summer days with abundant rainfall, sunlight and humidity. Some cultivars flower even under short day conditions.

By using growth regulators like GA3 flowering can be induced and also by removal of the tubers without damaging the main stem. Sterility has been reported due to absence of pollen grains; Similarly self and cross (unilateral & bilateral) incompatibility is observed in diploids and dihaploid forms. Tuberosum & andigena are freely crossable. Species within a series of crosses generally show cross compatibility. Outbreeding is a general rule in diploid tuber bearing Solanum species. The crossability barrier has been expressed as late as in F1 (Pushkarnath et al.,1965 & Nayar and Kishore, 1965).

Genetic Diversity and Genetic Resources

The variability was used from *Solanum tuberosum ssp. tuberosum* and fraction of variability was available from *ssp. andigena* which is a short day variety and is a rich source of resistance to diseases and good keeping quality. Cross between *tuberosum* and *andigena* gives more heterosis. Other species can also be used but restricted to for using resistant genes for Root Knot Nematode, Leaf Roll Virus and field resistance to Late blight etc.

Potato Classification

1. Origin and response to day length

Distinct character	*S. tuberosum* ssp. *tuberosum*	*S. tuberosum* ssp. *andigena*
Origin	Chile region	Andean region
Day length response	Long day plants	Short day plants
Polyploidy	Many types of polypolids	Tetraploids
Varieties	Uptodate, Magnum bonum, etc.	All the desi varieties Darjeeling Red Round, Phulwa and Gola

2. Based on crop duration

Character	Early potato	Late potato
Stolon length	Short	Long
Stem	Hollow	Solid
Stolon number	Less number	More number
Tuber colour	White	Coloured
Photoperiod	Long days for tuber development	Short day for tuber development
Varieties	Gola Kufri Chandramukhi	Phulwa Kufri Kisan

S. tuberosum has all the characters of early potato & *S. andigena* has all the characters of late potato. These two species are crossable. Many varieties developed are natural/artificial crosses.

Classification of Varieties Based on Maturity Groups

Early : (2-2 ½ month from sowing to harvest), Gola Up-to-Date, Kufri Chandramukhi, Kufri Chamatkar, K. Alankar

Mid : (2 ½ - 4 ½ months), Kufri Sindhuri

Late : (4 ½ - 5 ½ months), Phulwa, Kufri Kisan, K. Jeevan

Zones, Area Covered and Varieties Required

1. Western Indo-Gangetic plains: It includes Punjab, Haryana & Rajasthan. Varieties for this region should be early, moderately resistant to late blight, frost & heat tolerant, slow rate of degeneration & photoinsensitive.
2. Central Indo-Gangetic plains: Western U.P, M.P. requirements are medium maturity, moderately resistant to late blight, slow rate of degeneration.
3. Eastern Indo-Gangetic plains: Parts of U.P, Bihar, Bengal, Orissa & Assam plains. Short duration varieties, moderately resistant to late blight, slow rate of degeneration, red skinned & small sized potatoes.
4. Central plateau region and southern regions: Medium maturity, moderately resistant to late blight, slow rate of degeneration and adaptation to high temperature are the requirements.

Varieties /sources for good adaptation

Long day and temperate conditions: CP1412, CP2099.

Sub tropical, short day conditions: CP1475, CP1515.

Tropical and near equinox: CP1798, CP1824.

Resistant sources to biotic stresses

Late blight: CP1353, CP1412, CP2137 (Kufri Alankar).

Wart: CP2291, CP2307.

Common scab: CP2213, CP2263.

Potato tuber moth: CP1824, CP1832.

Tolerance to abiotic stresses

Heat tolerant lines: CP2108, CP2109, CP2150 (Kufri Laukar).

Soil salinity: CP1532, CP2061.

Tuber quality traits

High dry matter: CP1456, CP2142 (Kufri Dewa), CP2153 (Kufri Red).

High protein content: CP1404, CP2141 (Kufri Chandramukhi), CP2158 (Kufri Sindhuri).

High vitamin 'C' content: JEX/A-63, JEX/A-68, JEX/A-208.

Wild species tolerant to different biotic stresses are

Late blight: *Solanum bulbocastanum, Solanum demissum.*

Bacterial wilt: *Solanum microdontum.*

Potato Virus 'X', PVY: *Solanum chacoense, Solanum acaule.*

*Ap*hids: *Solanum berthanttii.*

Nematodes:

Cyst nematodes: *Solanum spegazzinii.*

Root knot nematode: *Solanum spegazzinii, Solanum microdontum.*

Genetic Resources at CPRI

Solanum tuberosum ssp. *tuberosum* : At CPRI by the end of 20th century totally there were 138 indigenous varieties (out of which, 24 cultivars are bred at CPRI, 11 old indigenous cultivars, 7 parental lines and 96 recently collected accessions) and 854 exotic collections from 32 different countries.

Solanum tuberosum ssp. andigena : 872 accessions from seven countries. 250 are wild, semi cultivated species from four countries.

***Tuberosum* species has narrow genetic base compared to *andigena* because** total time for evolution of *tuberosum* species is shorter than *andigena* and lot of accessions belonging to *tuberosum* are lost or extinct due to the blight epidemic of 1840's which reduced large amounts of variations in potato (Grun, 1990 and Hawker, 1990). *Andigena* species is a short day adapted and Andean potato is the ancestor of *tuberosum* has a wider genetic base. Indian potato varieties resemble more to *andigena* than *tuberosum. Tuberosum* have been developed in Europe (temperate regions) and introduced to India by British rulers. *Andigena* is a short day species and potatoes grown in India are also short day and are heterotic in yield. Hence, Indian potatoes are evolved by both *andigena* and *tuberosum* potatoes.

Some of the important features of Andean potato are tolerance to biotic and abiotic stresses, high starch and protein content, good keeping quality and responsiveness to fertilizer application is very low. If *andigena* is crossed with *tuberosum* the amount of heterosis is much more. Some of the Andean potato varieties in India are Darjeeling Red Round and Phulwa.

Genetics

Character	Number of genes	Type of gene action
Skin colour	Digenic (D-R-)	Complementary D_R_ : Red D_rr : White dd R_ : White dd rr : White
Position of eye	Monogenic	Flat eye is incompletely dominant to deep eye
Flesh colour	Monogenic	Deep yellow colour is incompletely dominant to white, the intermediate is pale yellow colour.
Stolon length	Monogenic	Long stolon is dominant to short stolon
Tuber shape	Monogenic	Long tuber axis is dominant to short axis

Breeding Objectives

1. High yield.
2. Insensitivity to photoperiod.
3. Good keeping quality i.e. resistance to shrinkage, rotting, tolerance against accumulation of sugars and should have reasonable amount of dormancy.
4. Responsiveness to fertilizers at varied levels especially at higher level.
5. Resistance to late blight: Earlier confined to colder climate of hills, now it has spread to plains also.
6. Slow rate of degeneration i.e. it should have high resistance to virus accumulation and MLO's which reduces the yield generation after generation.

   PVX- mechanical transmission.

   PVY, PVA, PLRV- aphid transmitted.
7. Tuber quality and suitable for processing. Quality refers to round medium sized tubers with shallow eyes and low sugar contents specially reducing sugars for chip making.

Breeding Methods

Introduction : Earlier only introduced varieties were ruling.

Primary introduction : Magnum Bonum, Craigs Defiance and Up-To-date.

Secondary introduction : Hybrid DN-45-developed through a cross between Katahdin x President. Kufri Kisan is a multicross derivative which is obtained by a cross between Ekishraju from Japan, Katahdin from U.S.A., Up-To-Date from Scotland and Phulwa an indigenous variety.

Clonal selection : Large number of clones are collected, evaluated and released as a variety. Kufri Red is a clonal selection from Darjeeling Red Round. K-1241 is a clonal selection from Phulwa.

Hybridization and selection : The flowers are borne in compound cymes terminally. Pentamerous and 2 celled ovary, single styled, bilobed stigma. Fruit is a berry. Fruit is commonly known as seed lot or apple which is 3cm in diameter. Flowers open mostly early in the morning. Self pollination is a rule, cross pollination occurs through wind, insects. Pollen germination completes with in 30 minutes, fertilization occurs in 12hours. Flowering and fruit set occurs in long day and low temperatures.

Kufri Kundan : Selection from a cross between Ekishrazu x Katahdin

Kufri Jyothi : Selection from a cross between A-3069 x A 2814.

Heterosis breeding: Heterosis is observed for earliness, tuber yield, tuber size, tuber weight. So in F1 seed production hand emasculation and pollination is followed after identification of good specific combiners. Pollen sterility is common. Few produce viable pollen but setting is very poor. Inbreeding depression is more. Seed set is poor and not exploited.

Back cross breeding : commonly used to transfer genes from wild species.

Interspecific hybridization : Other species are use full sources of biotic and abiotic stresses and other novel attributes.

Potato species resistant to biotic and abiotic stresses

Species	Utility
Solanum demissum lindl (2n = 72)	Resistance to late blight, virus A, virus Y
S. stoloniferum (2n = 48)	Resistance to virus x, frost, Colorado beetle.
S. vernei Bitt et,. Wittm (2n = 24)	Resistant to two species of Heterodera nematode.
S. multidissectum Waek (2n=24)	Resistant to two species of Heterodera nematode.
S. antipovczii Bulk	Resistance to late blight
S.curtilobum jug. Et. Buk	Resistant to frost
S . phureja jug. Et. Buk	Non dormant type used inbreeding
S . chacoense bitter	Non dormant type used in breeding Tolerence to high temperature
S . anomalocalyx, S. jamessi Torr; *S. Saltense*	Resistance to early blight.

Success can be achieved by

1. Using amphidiploids: Transfer of genes from one species to another.

   Diploid x Diploid
   ↓
   F1 (sterile)
   ↓ (Doubling of chromosomes)
   Amphi diploids

2. Using bridge species:

   a. *S. demissum* x *S. tuberosum*
   2 n = 72 ↓ 2 n = 48
   F1 sterile (univalents)

   S. demissum x *S. rybinii*
   (+LB,PVA,PVY) (Bridge species)
   ↓
   F1 (Fertile) x *S. tuberosum*
   ↓
   Fertile progenies

   b. *S. acuale* x *S. tuberosum*
   ↓
   F1 sterile (univalents)

   S. acuale x *S. simplicifolium*
   (+ PVX,PLV, PSTD, Phytoplasm) ↓ (Bridge species)
   F1 (fertile) x *S. tuberosum*
   ↓
   Fertile progenies

   Here *Solanum simplicifolium* and *Solanum rybinii* are used as bridging species to get fertile progenies.

3. Mixed pollination: using the pollen from both the parents i.e. from male and female parents.

4. Spray IAA and then go for pollination.

5. Grafting of female parent on male parent and then pollinate the scion with the pollen of the male parent.

Kufri Kuber: It is an interspecific hybrid from a cross between *Solanum curtilobum* and *Solanum tuberosum* followed by selection.

Polyploid breeding: Homozygous parental lines are produced in potato through development of haploids and then their polyplodization. Haploids can be produced by crossing of tetraploid variety with pollen from diploid variety.

Achievements

Varieties for Indo-Gangetic Plains: Kufri Kundan (1958), Kufri Kuber (1959), Kufri Red (1959), Kufri Safed (1958), Kufri Sindhur (1967), Kufri Alankar (1968), Kufri Chamatkar, Kufri Chandramukhi, Kufri Sheetman (1968), Kufri Badshah (1969), Kufri Bahar, Kufri Lalima (1982).

Varieties for Tarai region: Kufri Dewa (1974).

Varieties for Plateau region: Kufri Laukar (1972).

Varieties for Northern hills: Kufri Kumar (1958), Kufri Kundan (1958), Kufri Jeevan (1958), Kufri Jyothi (1968).

Varieties for Southern hills: Kufri Neela (1966), Kufri Neelamani (1968), Kufri Muthu (1971), Kufri Swarna (1985).

Varieties for Eastern hills: Kufri Khasigaro (1968), Kufri Naveen (1968), Kufri Sherpa (1983), Kufri Megha (1989).

From U.P, T.N, J&K Dept. of Agriculture some varieties were released: PS55, ON1645, K122, G45.

Significant Traits of Some of the Varieties

Kufri Sindhuri: It is a selection from a cross between Kufri Red x Kufri Kundan. It is resistant to LB, frost, 120 days to maturity and tubers are red coloured.

Kufri Sakthi: It is a sister clone of Kufri Red x Kufri Kundan.

Kufri Badshah (JF4870): It carries set of 'R' genes which has horizontal resistance to late blight.

Kufri Sheetman: It is a frost tolerant variety.

Kufri Khasigaro : Interspecific hybridization.

Hybrid 2236 did well at short and long day conditions.

HB827, HB841 and HB 858 suited for high temperature.

JW160: Tolerant to late blight with inter adaptability.

H92/621: Tolerant to heat and drought and resistant to mites and leaf hopper.

Most of the improved potato varieties are developed at the Central Potato Research Institute (CPRI), Kufri, Shimla.

Majority of early varieties are clonal selections derived from foreign varieties or from popular acclimatized varieties that survived after introduction. Varieties are distinguished based on their habit, pigmentation of stem, leaf, flower and shape, size, colour and depth of eyes and flesh colour of tubers. A large number of earlier varieties either developed at CPRI, Shimla or evolved as famous selections from improved varieties and were replaced by new improved varieties.

Satha, Gola, Up-to-date, Phulva, Great Scot, President, Kufri Kuber, Kufri Kisan, Kufri Neela, Kufri Khasi-Garo, Kufri Naveen, Kufri Chamatkar, Kufri Neelmani, Kufri Sheetman, Kufri Alankar, Kufri Jeevan, Kufri Moti, Kufri Lavkar and Kufri Dewa are a few of the earlier varieties.

Early Maturity Varieties

Kufri Chandramukhi : This is an early-maturing (80-90 days) variety and suitable for cultivation in Bihar, Gujarat, Haryana, parts of Himachal Pradesh, Karnataka, Madhya Pradesh, Maharashtra, Orissa, Punjab, Uttar Pradesh and West Bengal. It produces around 20-25 tonness/ha tubers. The tubers are large, white, oval, slightly flattened with fleet eyes and dull white flesh. This variety possesses medium tuber dormancy and is tastiest as vegetable. It is suitable for preparations of flakes, flour, chips, French fries and dehydrated products, when grown under warmer areas of central plains.

Kufri Jawahar: An early-maturing (80-90 days), it is ideal for cultivation in Haryana, Punjab, plateau region of Gujarat, Karnataka and Madhya Pradesh. It can produce 25-30 tonnes/ha tubers. The tubers are medium, white, round-oval with flat eyes and pale-yellow flesh.

Kufri Khyati: This is early-maturing (70-80 days) variety for cultivation in North Indian plains. It can produce 25-30 tonnes/ha tubers. The tubers are medium, white, round-oval with medium eyes and white flesh.

Kufri Ashoka: An early-maturing (70-80 days), it is suited for cultivation in Bihar, Haryana, Punjab, Uttar Pradesh and West Bengal. It produces around 20-25 tonnes/ha tubers. Tubers are medium-large, white, oval with medium-deep eyes and white flesh. Though the variety is susceptible to late blight but escapes late blight due to early maturity.

Kufri Surya: An early-maturing (70-80 days) and suitable for northwestern plains and rabi and kharif seasons. It can produce about 20-25 tonnes/ha tubers. The tubers are medium to large, white, oblong with shallow eyes and pale-yellow flesh. This is a heat tolerant variety and can be grown in areas having night temperature above 20°C and is field resistant to hoppers. Due to large oblong tubers and high dry-matter content it can be used for preparation of French fries.

Kufri Lauvkar: This is early-maturing (75-80 days) variety and suitable for cultivation in Plateau region of Karnataka, Madhya Pradesh and Maharashtra. It can be grown in kharif and rabi seasons. It produces 25-30 tonnes/ha tubers. It is able for rapid bulking under warmer conditions. Tubers are large to medium, white, round with fleet eyes having prominent eyebrows and white flesh. This is suitable for preparation of flakes, flour, chips and dehydrated products when grown in warmer areas of central plains.

Medium Maturity Varieties

Kufri Jyoti: It is a medium-maturing (90-100 days) for cultivation in North and South Indian hills. It produces around 25-30 tonnes/ha tuber yield. The tubers are medium, white, oval with fleet eyes and white flesh. The variety is moderately resistant to early blight and immune to wart and has slow rate of degeneration. It is suitable for flakes, flour, chips and dehydrated products when grown in warmer areas of central plains.

Kufri Bahar: This is medium-maturing (90-100 days) variety for cultivation in parts of Haryana, Madhya Pradesh, Jammu and Kashmir and Uttar Pradesh. Now this is the prominent variety of Uttar Pradesh, producing 30-35 tonnes/ha tubers. It can also be grown in early planting season. The tubers are large, white, oval with medium deep eyes and white flesh.

Kufri Lalima: It is a medium-maturing (90-100 days) for cultivation in Bihar and Uttar Pradesh. It is popular in north-eastern plains having preference of red-skinned tubers. The variety is capable of producing 30-35 tonnes/ha tubers. The tubers are medium large, red, round with medium deep eyes and white flesh. The variety is moderately resistant to early blight and resistant to PVY.

Kufri Sutlej: It is a medium-maturing (90-100 days) variety for cultivation in Bihar, Haryana, Madhya Pradesh, Punjab and Uttar Pradesh. It produces around 35-40 tonnes/ha tubers. Tubers are attractive, large, white, oval with fleet eyes and white flesh. The variety is moderately resistant to late blight.

Kufri Pukhraj: This an early-bulking/medium maturing (80-90 days) variety for cultivation in Bihar, Gujarat, Haryana, Himachal Pradesh, Madhya Pradesh, Karnataka, Maharashtra, Orissa, Punjab, Uttar Pradesh and West Bengal. It can produce 3 5-40 tonnes/ha tubers. Tubers are large, white, oval, slightly tapered with fleet eyes and yellow flesh. The variety is resistant to early blight and moderately resistant to late blight.

Kufri Anand: This is medium-maturing (90-100 days) variety for cultivation in plains of Uttar Pradesh and other states. It is high-yielding variety and can produce 35-40 tonnes/ha tubers. Tubers are large, white, oval-oblong flattened with fleet eyes and white flesh. The variety is resistant to late blight, tolerant to frost and can be used for preparation for French fries.

Kufri Sadabahar: This is a medium-maturing (90-100 days) variety for cultivation in Uttar Pradesh. It can be grown in adjoining states as well and is capable of yielding 35-40 tonnes/ha tubers. The tubers are medium to large, white, oval-oblong with shallow eyes and white flesh. The variety possesses field resistance to late blight and tolerance to frost. Due to large oblong tubers and high dry matter, it can be used for preparation of French fries.

Kufri Arun: It is a medium-maturing (90-100 days) variety for cultivation in North Indian plains. The variety can produce 35-40 tonnes/ha tubers, possessing about 20% dry matter. The tubers are medium, red, oval with shallow-medium eyes and creamy flesh. The variety is field resistant to late blight and tolerant to frost.

Late Maturing Varieties

Kufri Badshah: It matures in 100-110 days. It is suited for cultivation in parts of Gujarat, Haryana, Jammu and Kashmir, Madhya Pradesh, Punjab and Uttar Pradesh. It is capable of producing 35-40 tonnes/ha tubers. The tubers are large, white and oval with fleet eyes and dull white flesh. The variety possesses moderate resistance to late blight, early blight and potato virus X.

Kufri Sindhuri: The late-maturing (100-110 days) variety Kufri Sindhuri is ideal for cultivation in Bihar, Gujarat, Karnatka,Jammu and Kashmir, Maharashtra, Punjab and Uttar Pradesh. It yields 35-40 tonnes/ha tubers. Tubers are medium, red, round with medium deep eyes and dull white flesh. It is moderately resistant to early blight, tolerant to PLRV and has slow rate of degeneration and possesses long tuber dormancy. It is also tolerant to temperature and water stress. It is suitable for dehydrated instant flakes and canning.

Kufri Chipson 2: It has round tubers with shallow eyes and creamy flesh. The variety is well adapted in North Indian plains. It can be grown in main autumn season. Its tubers have average dry-matter content of 21-24% and reducing sugar of 30-100 mg/100 g of fresh weight depending upon location and cultural practices. The variety achieves maturity in 110-120 days. This is a very good variety for chips and flakes making. Both Kufri Chipsona 1 and Kufri Chipsona 2 have low phenolic content and thus posses less problem of enzymatic browning. The keeping quality is inferior to Kufri Chipsona 1. The chemically mature tubers yield acceptable coloured chips for 4-5 months during storage at 10-12°C.

Kufri Chipsona 1: It is suitable for making chips, French fries and flakes. This variety can be grown in North Indian plains in main autumn season, with an yield potential of 300-350 q/ha. Its tubers have average dry-matter content of 20-23% and reducing sugars of 10-100mg/100 g of fresh weight depending upon the location and cultural practices. The tubers yield acceptable chip colour from 90 days maturity onwards. The 110 days grown crop keeps well at 10-12°C for 4-5 months with respect to chipping quality. The variety has excellent keeping quality.

Kufri Chipsona 3: Its tubers are of ovoid shape with medium deep eyes and creamy flesh. The total and processing grade tuber yield is higher than Kufri Chipsona 1 and Kufri Chipsona 2. This variety is resistant to late blight with an average yield potential of 300-350 q/ha. The chemical maturity of variety reaches in 110 days and tubers yield acceptable chip colour for 4-6 months during storage at 10-12°C. Kufri Chipsona 3 has good processing quality with more than 20 % dry-matter content and very low reducing sugars (< 0.05 %) in plains.

5

Okra (Ladies Finger)

Okra has higher average nutritive value (ANV) of 3.21 which is higher than tomato, egg plant and most of the cucurbits except bitter gourd (Grubban, 1977). Its roasted seeds are substitute for coffee. It is one of the export potential crop where it accounts to 60 per cent of fresh vegetables (excluding potato, onion and garlic) exported from India. Importing countries are Middle East, Western Europe and USA.

Origin and Distribution : It belongs to the genus *Abelmoschus* established by Medikus, 1787, however de Candolle (1824), treated it as section of *Hibiscus*. In 1924 Hochreuntiner reinstated the genus *Abelmoschus* of Medikus stating that calyx, corolla and stamen fused together at the base and fall as one piece after anthesis (deciduous) where as in case of hibiscus these are persistent. Though the genus is of Asiatic origin, the origin of cultigen *Abelmoschus esculentus* is variable like India (Master, 1875) Ethiopia {de Candolle (1883) and Vavilov (1957)}, Western Africa (Chevalier, 1940 and Mardorct, 1947) and Tropical Asia (Grubban, 1997) and probably its might have entered India at the end of 19th century. Cultivated okra is of old world origin. It arose as a cultigen in the Abyssinian region (Ethiopia). Wild species are indigenous to Africa. Three wild species *Abelmoschus tuberculatus, A. moschatus* and *A. ficulneus* are native to India. Introduced to Mediterranean region to Europe and Asia. But during Columbian period it was introduced to America. It still turnout to be polyphyletic (many places) in origin. *A. tuberculatus* (Indian origin) and an unknown species cross resulted into *A. esculentus.*

Chromose number*:* Gametic chromosome number ranges from 59 to 72. Present Indian varieties have n = 65.

Genetic Resources: NBPGR- has both indigenous and exotic collections. IPGRI has designated NBPGR with global responsiblity of base collection of okra. Besides IARI, IIVR, IIHR and Agricultural universities are having okra collections.

Cytogenetics: Cultivated types are *Abelmoschus esculentus* and *Abelmoschus manihot* ssp *manihot* are under cultivation in limited scale in Africa. *Abelmoschus esculentus is* polyploidy in nature (Joshi and Harden, 1956). *Abelmoschus esculentus* has 2n =130 and it is postulated that this species is evolved by hybridization between n=29 (Seems to be *Abelmoschus tuberulatus* 2n=58) and n=36(unknown species) followed by doubling of chromosome number. This theory is postulated since on crossing of *Abelmoschus tuberculatus* with *Abelmoschus esculentus* resulted with 29 bivalents and 36 univalents. Second progenitor still remain undiscovered. Other species and their chromosome numbers are *Abelmoschus manihot (2n= 68), Abelmoschus tuberulatus (2n=58), Abelmoschus ficulatus (2n=72), Abelmoschus moschatus(2n=36)* and *Abelmoschus tetraphyllyus (2n=130)*. *Abelmoschus moschatus* X *Abelmoschus esculentus* progenies retrieved through invitro embryo culture little genomic affinity in that 8 are bivalent and 84 are univalent.

Genetic Studies

Okra is often cross-pollinated crop where protogyny is reported in few varieties. Early germination indicates early types. Number of primary branches and plant height positively associated with yield.

Pod yield positively correlated with seed yield. Flower position and number of anthers and number of pollens per flower.

Flower position/ node number	Number of anthers	Number of pollen grains
1	26	1164
2	68	4828
4	57	4161
5	43	2365
10	10	450

High heritability and high genetic advances reported for fruit diameter, fruit length, crude fiber, total sugar and Vitamin-C content (Malik, 1968), plant height and number of days to flowering (Rao, 1972).

Plant height: a) Monogenic, tall dominant to dwarf. b) 4-5 groups of dominant genes. Earliness and more number of fruits: Dominant and over dominant and 1-3 groups of dominant genes also reported. Leaf margin is monogenically controlled where cut leaves dominant to entire (non lobed) leaves. Pod colour is also monogenically controlled where white is dominant to green pod. Spininess is monogenic where spiny is dominant to non spiny. Fruit hairiness is monogenic where hairiness is completely dominant to non hairiness. Stem colour is monogenic where purple stem colour is dominant to green. Pod shape is oleigogenically controlled where angular shape is dominant over round and epistatis has been observed. Yield, branches and days to flower are controlled by additive genes. Predominance of additive gene effects for days to flowering, dominant gene effects for plant height and additive and dominant both for number of fruits per plant have been reported (Rao, 1977). Rao and Ramu (1977) reported additive gene effects for days to flowering, number of pods and yield and Non additive gene effects for plant height and number of seeds per pod. Kulkarni *et al.*, 1976 reported additive gene action for number of fruits and non additive gene action for days to flowering and plant height. Arora, 1980 reported additive effect for days to flowering, length and weight of fruit, seed weight, number of ridges, fruits per plant, nodes per plant, days to marketable maturity, total yield and protein content. Venkataramani (1952) reported that dark green colour is dominant over light green, while green is recessive to greenish red. Kolhe and Dcarz, (1966) reported monogenic control of pigmentation of calyx, corolla and fruit. Fruit hairiness and leaf lobing are controlled by single dominant gene and fruit colour had a digenic complimentary control (Nath and Dutta, 1970).

Genetics of YVMV Resistance

1. Singh *et al.* (*1962*): IC1542: two recessive alleles of two different loci: yv1, yv2:
2. Thakur (1976): *A. manihot* ssp *manihot* cv. Ghana: Two complementary dominant genes, environment sensitive and polygenic resistance can not be ruled out.
3. Jhamble and Nerkar: Single dominant gene in *A. manihot and A. manihot* ssp *manihot*

Objectives of Breeding

1. High marketable yield (dark green fruits, tender, thin, smooth and 4-5 ridges on pod).
2. Early maturity with prolonged harvest period.
3. Evolve the varieties resistant to YVMV and also *Fusarium* wilt, *Cercospora* leaf spot, powdery mildew, *Alternaria* leaf spot, fruit rot, anthracnose, etc.
4. To combine the resistance to YVMV with resistance to fruit and shoot borer, white fly, jassids and root knot nematodes.
5. Develop multiple disease and pest resistant varieties.
6. Development of suitable ideotypes: short plant, more nodes, devoid of conspicuous hairs and easy separation of fruits.

7. Varieties with optimum seed setting ability.
8. Tolerance to low temperature, excess rain, salinity and alkaline soil.
9. Development of export type.
10. Development of processing type varieties (Dehydrated).

Important centers working

PAU, IIHR, TNAU, Division of plant introduction (IARI), GAU, APAU, U.P., IIVR, Varanasi, etc.

Breeding Methods

Pure line selection, mass selection, backcross method, pedigree method, bulk population method are adopted and suggested.

Introduction

Ghana is the variety: From Africa known as *Abelmoschus manihot* ssp *manihot* introduced to India, which is source of resistance to YVMV.

Prue line selection

Pusa Makhmali: Bred from natural collection from West Bengal. **Co-1** is the single plant selection from Red Wonder.

Pedigree method: Pusa Sawani, Punjab Padmini, Parbhani Kranti, P7, Arka Anamilka and Arka Abhaya are varieties developed where principles of Pedigree Method involved.

Mutation breeding: Mutation breeding was attempted to evolve early yielding varieties resistant to yellow vein mosaic virus.

Polyploid breeding

- Polyploids showed inhibited growth and retarded flowering.
- Failure of many species crosses due to hybrid sterility prompted many workers to produce amphidiploids.
- *A. esculetus* (2n = 130) x *A. manihot* ss. *manihot* (2n = 194)

  n=65 ↓ n=97

  F_1 was sterile (7.07% seed fertility)

  2n=162 = 65 + 97 univalents

  Colchicine treatment ↓

  Amphidiploids 2n = 324 162 bivalents

Obtained amphidiploids which were fertile (81.8% seed fertility) and carried resistance to YVMV.

- *A. esculentus* (n = 65) x *A. manihot ssp. tetraphyllus* (n = 69)
  2n=130 ↓ 2n=138
  F1 was sterile (7% seed fertility)
  2n=134 = 65 + 69 univalents
  Colchicine treatment ↓
  Amphidiploids 2n= 268 [134 bivalents]

Obtained amphidiploids which were fertile (94.4% pollen stainability) and carried resistance to YVMV

Heterosis: Heterosis is reported for fruits size, fruit weight and No. of fruits/ plant in okra. (Warrier, 1946 and Venkataramani, 1952) and for plant height, fruit size, no of fruits and yield (Joshi *et al.*, 1958). Yield hetrosis varied from 11.8% to 101.7% (Sharma, 1965). It is exploited to very little extent by private companies (hand pollination and emasculation). There are no F1 hybrids released from public sector institutes in India and it can be attributed to lack of economically viable seed production system and increased fruit size has less market preferance.

Inter specific hybridization: Attempted between (Mamidwar *et al.*, 1979) *Abelmoschus esculents* and *Abelmoschus tetraphyllyus* Roxb. *(n=69)*, *Abelmoschus manihot* Medik *(n=33)* and *Abelmoschus manihot* (L) Medik *ssp manihot (n=97).*

1. *Abelmoschus esculents* and *Abelmoschus tetraphyllyus:* Chromosomes resembling was there but there was total seed sterility in the hybrid, but still little success in nature lead to introgression.
2. *Abelmoschus esculentus(n=65)* and *Abelmoschus manihot (n=33):* Resulted with 37^{II} (bivalents) and 55^{I} (univalents) (Jambhale and Nerkar, 1982) and bivalents ranged from 20 to 40 (fruit without seed). Induced amphidiploids to overcome sterility and resulted with $129\text{-}134^{II}$
3. *Abelmoschus esculents* X *Abelmoschus manihot ssp manihot:*
   (2n = 130) ↓ → (2n=194)
   (7 % of F_1 fertile) →
   162^{II} Colchicine induced amphidiploids (88.2% fertility)
4. Suresh Babu (1987)
   Abelmoschus esculentus X *Abelmoschus manihot ssp tetraphyllyus* var tetraphyllyus
   (2n = 130) ↓ → (2n = 194)
   (2n = 134) → 36^{II} & 62^{I}

Sterility of F_1 is attributed to failure of development of female gametes. Induced amphidiploids showed regular meiosis forming 134^{II} (2n=268) and pollen stainability of 94%.

5. Partial success when high chromosome number spp used as a female otherwise no success. (between - *Abelmoschus esculents* X *Abelmoschus manihot*).
6. Genes for YVMV have been successfully transferred from
   i. *Abelmoschus manihot ssp manihot* to *Abelmoschus esculents.* (Thakur-1976, Singh and Thakur- 1979, Jambhale and Nerkar-1981, Dhilon and Sharma-1982)
   ii. *Abelmoschus manihot ssp tetraphyllyus* to *Abelmoschus esculents* (Dutta, 1982)

Disease Resistance

Due to YVMV, 50% to 90% yield loss reported

Fusarum and related wilts (north India and other parts also, spring season)

Cercosporsa leaf spot –recently reported to be serious.

Powdery mildew, Alternaria leaf spot, fruit rot, anthrocnose, etc. sporadically reported but can be controlled by fungicides quite effectively.

YVMV

IC1542 - Earliest collection from West Bengal and developed Pusa Sawani: symptomless carrier.

Sadhu *et al.* (1974) identified the accession EC31830, Asuntem Koko from Ghana, identified as *Abelmoschus manihot* Medicus ssp *manihot* by Royal Botanical Garden, Kew London was almost immune to YVMV.

It has contributed for development of

- Punjab Padmini
- Punjab-7
- Parbhani Kranti

Sources

1. *Abelmoschus manihot* (L) *medicus*- two accession highly resistance viruses (Arumugan *et al.*, 1975)
2. *Abelmoschus manihot var.* Pungen
3. *Abelmoschus crinitus*
4. Abelmoschus tetraphyllus
5. *Abelmoschus manihot* (L) Medicus

(3–4: Highly resistance to YVMV Nariani and Seth, 1958)

6. *Abelmoschus manihot* (L) Medicus ssp *manihot*
7. *Abelmoschus manihot* ssp *tetraphyllus*
   ↓ (Ugale *et al.*, 1976 and Mamidwar *et al.*, 1979)

   Nerkar and Jambhaale, 1985

   Successfully used to development of Arka Abhay and Arka Anamika (Dutta, 1984).

Genetics

1. **Singh *et al.*,** 1962 - two recessive allele of two different loci

   IC1542 - YV_1 / YV_1, YV_2 / YV_2

   Pusa Makhmali - YV1/YV1, YV2/YV2
2. **Thakur** (1976) - *Abelmoschus manihot ssp manihot* cv. Ghana

   Two complimentary dominant gene and these genes are sensitive to environmental changes especially low temperature, however polygenic resistance cannot be ruled at.
3. **Jambhale and Nerkar**, 1981 – single dominat gene in *Abelmoschus manihot* and *Abelmoschus manihot ssp manihot*
4. The inheritance pattern studies by X^2 test carried out by Pullaiah et al., 1998, suggested that resistance to YVMV was controlled by two complimentary genes in susceptible X susceptible crosses.

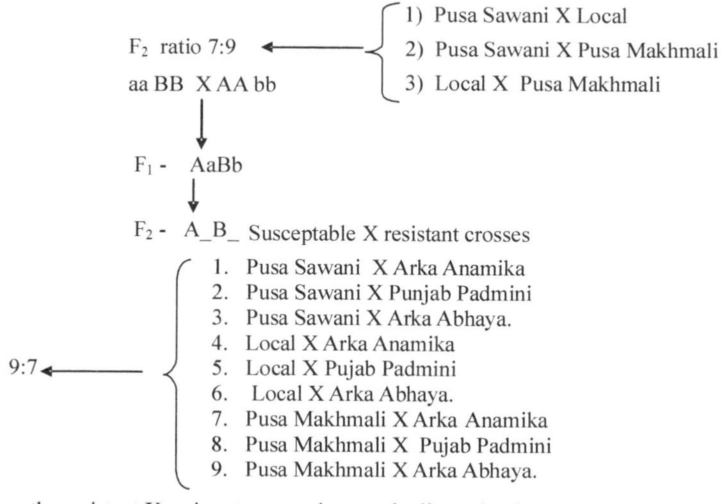

Where as the resistant X resistant crosses by two duplicate dominant genes

15:1 ← 1) Arka Anamika X Pujab Padmini; 2) Arka Anamika X Arka Abhaya.; 3) Pujab Padmini X Arka Abhaya

The genetic ratio of test crosses plant ⟶ 3:1 in all these crosses

Nerkar and Jambhale (1985) reported that the transfer of resistance from *Abelmoschus manihot* was successful by two back crosses followed by selection in selfed generations.

While that from *Abelmoschus maniho* ssp *manihot* was successful by growing straight generations. But crosses with *Abelmoschus tetraphyllus* met with failure. Resistance to YVMV in either of species was controlled by single dominant gene.

F_1 hybrid developed from - IIVR are DVR-1, DVR-2, DVR-3 and DVR-4.

- TNAU CO-3 field resistant to YVMV

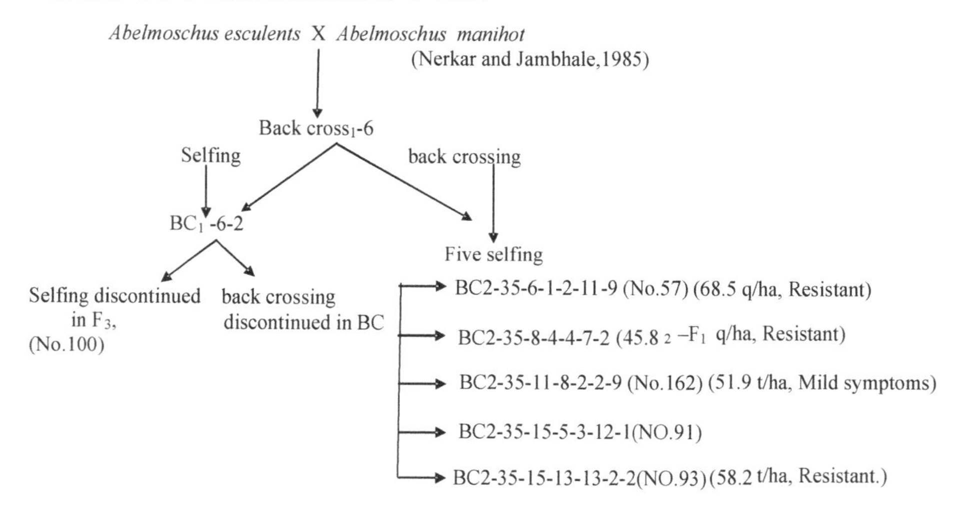

The genetics is not consistently reported and is not conformed because of non-availability of sterility in the F1.

Genotypes Nun 1145 and Nun 1144 showed moderate resistance reaction and symptoms appeared after 25–30 days of inoculation. Whereas, the genotypes Nun 1142, Nun 1140 and M10 showed moderately susceptible reaction and symptoms were produced after 15–25 days after the inoculation under artificial condition (Venkataravanappa *et al*., 2013).

The genotypes Nun 1145 and Nun 1144 showed moderate resistance and genotypes Nun 1140, Nun 1142, Nun 1143 and M10 showed moderately susceptible reaction under natural condition (Venkataravanappa *et al*., 2013).

The non-radioactive digoxigenin-labeled DNA probe was used in dot-blot hybridization to detect the virus in the total DNA isolated from the symptomatic plant as well as nonsymptomatic plant of different okra genotypes. The probe could be detected up to a concentration of 10-2 dilution in the plant showing the yellow vein mosaic disease (Venkataravanappa *et al*., 2013).

The mean incidence of OYVMV in summer season was low (22.25%) compared to high incidence in rainy seaon (51.95%). The intensity of OYVMV was very high in the rainy season crop due to the favouarable environmental condition for the whitefly and virus (Solankey *et al.*, 2014).

Sequencing analysis of the DNA-A component from various isolates collected from India revealed that 80% of the isolates aligned into a common cluster and virus isolates from Gujarat were highly distinct and aligned to two distinct clusters. The isolates with Enation like symptoms were nearly identical (95.5%) with Cotton Leaf Curl Gemini Virus and isolates with stem bending symptoms fall into distinct species. Rest of the isolates were identical to common Tomato Leaf Curl Virus. Under favorable environment (High temperature & low humidity), the vector (*Bemisia tabaci*) can carry 3 different viruses for 6 days simultaneously.

Amphidiplody, marker linkage, stability, and complexicity in copy number were the main drawbacks in using Marker Assisted Selection in okra breeding. The marker data gave correct interpretation in > 95% samples. Since the distribution of alleles is random there will be some population of plants with genome more similar to recurrent parent i.e genetic similarity >92.5%. Statistically we can get plants similar to BC2 in BC1 itself. Such plants with more genetic similarity can be taken for next cycle of backcrossing. The cycle of screening is continued and a breeding program can be completed within 3-4 cycles of back cross instead of traditional 7-8 cycles. For complex genomes like okra- a successful breeding program can be effectively completed faster by combined approach of MAS (Marker Assisted Selection) and Genotypic screening." (Dr. Kiran V Hegde of Chromous Seeds).

In Asia 21 plant viruses are reported and among these 19 viruses are reported in India. Twenty seven begomoviruses and one each of ilarvirus, nepovirus, potyvirus, tospovirus and tymovirus infect okra and cause several viral diseases. In tobacco streak virus infected plant, fruits become distorted and this disease is transmitted by thrips and seeds. Yellow vein mosaic is most devastating disease of okra causing > 60% yield loss. okra enation Leaf curl virus is serious in North. TSV an ilarvirus is an emerging problem for okra production. Begomoviruses have high recombination rate and the presence of B-biotype whiteflies are contributing to epidemics of begomoviruses in okra. Host genetic resistance to viruses is one of the most practical, economical and environmentally secures strategies for reducing yield losses in okra. Begomoviruses possess a genome comprised of one or two circular ss-DNA molecules, genome is monopartite (2.5–3.0 kb) or bipartite (2.5–5.0 kb), having smallest known genome for an independently replicating twinned icosahedral virions [18X30nm] and transmitted primarily by whitefly *Bemisia tabaci* and few are also transmitted mechanically. Natural resistance is available in wild species and land races of bhindi which has to be identified for each virus and ecological region. In India okra crop is infected by seventeen types of begomoviruses. Efficient Screening of genetic materials may be done through screening in hot spots/regions, artificial screening through whitefly and screening through grafting/agroinoculation. Resistance evaluation can be performed through sensitive assay like visual symptoms, ELISA/nucleic acid probe and PCR assay. Punjab, Haryana, Andhra Pradesh, Jharkhand, West Bengal and Odisha were identified as hot spots for yellow mosaic disease screening. Zero per cent disease incidence along with negative probe/PCR gives immune reaction, while 1-10% and 11-25% disease incidence with positive/negative and positives probe/PCR provide highly resistant and resistant, respectively (Dr. M. Krishna Reddy, IIHR, Bangalore).

Variety	Resistant to virus	Name of the Institute
Pusa sawani	YVMV	IARI, New Delhi
Pusa A-4	YVMV	
Selection-2	YVMV	
Arka Anamika	YVMV	IIHR, Bangalore
Arka Abhay	YVMV	
Kashi Bhairav (DVR-3)	YVMV &OLCV	IIVR, Varanasi
Kashi Mohini (VRO-3)	YVMV	
Kashi Vibhuthi (VRO-5)	YVMV &OLCV	
Kashi Mangali (VRO-4)	YVMV& OLCV	
Kashi Pragathi (VRO-6)	YVMV &OLCV	
Kashi Satdhari (IIVR10)	YVMV	
Sheetla Uphar (DVR-1)	YVMV	
Sheetla Jyothi (DVR-2)	YVMV & OLCV	
Punjab Padmini	YVMV	PAU, Ludhiana
Punjab-7	YVMV	
Punjab-8	YVMV	
Varsha Uphar	YVMV	HAU, Hisar
Hissar Naveen	YVMV	
Hissar Unnat	YVMV	
HBH-142	YVMV	
Azad Kranthi	YVMV	CSAUA&T, Kanpur
NDO-10	YVMV	NDUA&T, Faizabad
Parbhani Kranthi	YVMV	MPKV, Rahuri
Phule Utkarsha	YVMV	
Utkal Gauvrav	YVMV	OUAT, Bhubaneshwar
CO-1	YVMV	TNAU,Coimbatore
CO-2	YVMV	
Susthira (AE-286-1)	YVMV	KAU,Vellanikara

Fungal Diseases

1. *Fusarium oxysporum*-resistance sources are IS9273, IS9857, IS6653, IS7194, CS3232, Pusa Sawani, and Pusa Makamali (Grover and Singh, 1970)
2. Red Ghana, Sel.-7-1, BH-27, IC-12096, IC-17252 – resistance to damping off and *Rhizactonia solani.*
3. Nigeria, EC-32598, IC-8248 – resistance to powdery mildew (*Erishephe cichoracearum*)

Nematodes

1. *Meladogyni incognita*
   a) Abtalia- resistance source from Iraq, which was slightly susceptible (Mahajan and Sharma,1979)
2. *Meladogyni javanica* –Long Green Smooth as a resistance source (Birat, 1964)

Insect Resistance

1. **Cotton jassid** - IIHR-21 (Dutta, 1971) - *Abelmoschus moschatus* (Singh, 1988)
2. **Fruit and shoot borer**-Red-I, Red-II, Red Wonder-I, Red Wonder-II (Srinivasan and Narayanaswami, 1961)-*Abelmoschus manihot* (Chellaiah and Srinivasan, 1983)-AE-75, Pusa Sawani and Long Green. (Madan and Dumbre, 1985)

Varieties Developed

1. Pusa Makhmali: developed by H.B., Singh and S. M., Sikka at plant Introduction division, IARI, New Delhi ,1955.
2. CO-1: TNAU, Coimbatore, 1976, single plant selection from a population of Red Wonder collected from Hyderabad. It is field tolerance to YVMV but Susceptible to fruit borer and powder mildew
4. MDU-1: TNAU, Coimbatore, 1975, it is an induced mutation from Pusa Sawani
5. Punjab Padmini : Evolved by B.R. Sharma in 1982 at PAU, field resistance to YVMV, Jassids and cotton bole worm

   Abelmoschus esculentus cv. Rashmi *X A. manihot ssp manihot cv.* Ghana

   F1 hybridized with F2

   ↓ (*A. esculentus cv.* Pusa Sawani X *A. manihot* ssp *manihot* cv.Ghana)

   F8

Ludhiana Sel.-1 named as Punjab Padmini it is filed tolerance to YVMV, jassids and cotton bole worm

6. **Gujarat Bhendi**: GAU, Ahmedabad 1983, pure line selection from unknown bulk seed sample received from IARI, New Delhi
7. **Harbhajan**: Original "Perkins Long Green" or "Sel.-6" named as Harbhajan in the Memories of Dr. Harbhajan by Dr.T. A. Thomas and R. Prasad
8. **Sel. -2:** evolved by H. B. Singh *et al.,* 1973-74, tolerance to YVMVPusa Sawan Best-1 X (Pusa Sawani X IC-7194)
9. **Parbhani Kranti** : Evoleved by N. D. Jambhale and Y. S. Nerkar of Maratwada Agricultural University, Parbhani in 1983. It has field tolerance to PM & YVMV

Abelmoschus esculentus cv. Pusa Sawani X *Abelmoschus manihot*

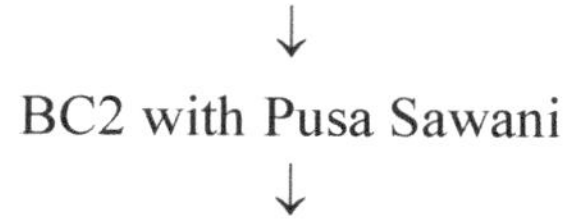

Selfing and selection up to F8

10. **P7:** Resistance to YVMV developed by M.R. Thakur and S.K. Arora in 1985 at PAU Ludhiana

Abelmoschus esculentus cv. Pusa Sawani X *Abelmoschus manihot* spp *manihot*

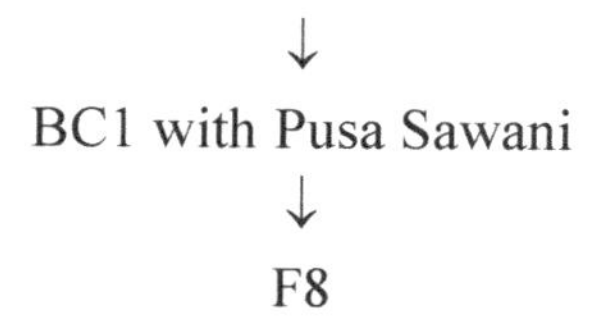

11. **EMS-8 :** Developed by B. R. Sharma and S.K. Arora,1989,PAU, Ludhiana. It is induced mutant from Pusa Sawani. 1% EMS, M8 generation final selection. It is field tolerance to YVMV and field tolerance to fruit borer.
12. **Pusa Sawani:** Resistant to YVMV, Now become susceptible to YVMV and it was developed by the selection in segregating generation of cross Pusa Makhmali x IC 1542 where, IC 1542 carried resistance to YVMV. It was developed by H. B., Singh, 1957-58, plant introduction division of Botany
13. **Arka Abhay** (IIHR-4): Resistant to YVMV and it was developed by the back cross and selection in segregating generation of cross *A. esculentus* (IIHR-20-31) X *A. manihot* ssp. *tetraphyllus* (Resistant to YVMV). Backcrossing was with *A. esculentus* (IIHR-20-31). Duration 120-130 days.

14. **Arka Anamika** (IIHR-10): Resistant to YVMV and it was developed by the back cross and selection in segregating generation of cross *A. esculentus* (IIHR–20-31) X *A. manihot* ssp. *tetraphyllus* (Resistant to YVMV). Developed through backcrossing with *A. esculentus*. Duration 130-135 days.

Other Varieties

Versha Upahar

Perkins Long Green, Sel. 2, Sel-6,

Pusa A-4: Resistant / tolerant to YVMV and shoot and fruit borer.

YVMV transmitted by white flies.

Varsha Uphar (HRB-9-2)

This is an YVMV resistant veriety, developed by CCSHAU, Hisar, derived from the cross Lam Selection-1 x Parbhani Kranti, Plants are medium-tall (90-120 cm), erect, short internode producing 2-3 branches with five lobed leaves. Fruits are dark green, 18-20 cm long, smooth with five ridge; first picking starts at about 45 days after sowing; gives an average fruit and seed yield of 9.8 t/ha and 1.7 t/ha respectively. This is popular among the farmers of Punjab, UP, Haryana and Delhi. Seeds could be obtained from the branches of NSC.

Pusa Makhmali

This is an early maturing variety developed by IARI, New Delhi; derived through selection from a local collection from West Bengal. Pods are smooth, straight, five ridged, attractive, light green, cylindrical and 15-20 cm long; gives yield of 8-9 t/ha. This is popular among the farmers of HP. Seeds could be obtained from the Division of Vegetable Crops, IARI, New Delhi.

Pusa Sawani

This is one of the oldest varieties of okra developed by IARI, New Delhi. Plants are tall, usually single stemmed and early maturing. Fruits are long, dark green and spineless; first picking starts on 55-60 days after sowing, susceptible to YVMV; gives an average yield of 12 t/ha. This is popular among the farmers of J&K, HP, Uttaranchal, North East states and Andman & Nicobar Island. Seeds could be obtained from Division of Vegetable Crops, IARI, New Delhi.

Punjab-7 (P-7)

This is an YVMV resistant variety developed by PAU. Ludhiana; derived through bacross method from an inter specific cross A. esculentus cv. Pusa Sawani x A. manihot sub, monihot cv. Ghana, Plants are medium-tall (85-105 cm) with short internodes, stem with pigmentation and deeply lobed leaves with less scrrated margins, basal protion of the petiole is deeply pigmented. Fruits are medium-long (15-20 cm), dark green, tender and 5-ridged, 15-21 g of weight, remain tender for 3-4 days; blunt tips and slightly furrowd; resistant to jassid and cotton ball worm; first picking starts on 54 days after sowing; gives an average yield of 9.5 t/ ha during rainy and 5 t/ha during sprin season; average seed yield of 0.4-0.8 t/ha. This is popular among the farmers of J&K, HP and Punjab. Seeds could be obtained from the Division of Vegetable Crops, PAU, Ludhiana.

Pusa A-1

This is an YVMV resistant variety, bred by IARI, New Delhi. Fruits are dark green and 12-15 cm long. First picking starts at about 45 days after sowing; gives an average yield of 14 t/ha; tolerant to aphids and jassids. This is popular among the farmers of UP, Bihar and Jharkhand. Seeds could be obtained from the branches of NSC.

GO-2

This variety has been developed by GAU, Anand. Fruit are green, tender and attractive. Resistant to YVMV under field conditions; gives an average yield of 10 t/ha. This is popular in Gujarat. Seeds could be obtained from the developing center.

Arka Anamika

An interspecific hybrid between Abelmoschus esculentus (IIHRE20-31) x *A. manihot* spp. tetraphyllus (Res. to YVMV) followed by backcross.

Plants tall well branched. Fruits lush green, tender and long.

Fruit borne in two flushes. Purple pigment present on both sides of the petal base.

Green stem with purple shade.

Fruits free from spines having 5-6 ridges, delicate aroma.

Good keeping and cooking qualities.

Resistant to Yellow vein mosaic virus Duration 130-135 days.

Yield 20 t/ha.

Arka Abhay

- An interspecific hybrid between Abelmoschus esculentus (IIHR 20-31) x *A.manihot* spp.
- Tetraphyllus (Res. To YVMV) followed by backcross. Plants tall, well branched.
- Fruits lush green, tender and long. Fruits borne in two flushes.

- Purple pigment present on both sides of the petal base.
- Green stem with purple shade. Fruits free from spines having delicate aroma.
- Good keeping and cooking qualities.
- Resistant to yellow vein mosaic virus (YVMV) Duration 120-130 days. Yield 18 t/h

Hybrid- Kashi Bhairav

Plants of this hybrid are medium tall with 2-3 branches; fruits are dark green with 10-12 cm length at marketable stage; yield 200-220 q/ha. This is resistant to YVMV and OLCV under field conditions. This has been released and notified during the XII meeting of Central Sub Committee on Crop Standard Notification and Release of Varieties for Horticultural Crops for the cultivation in the entire okra growing region of the country.

Kashi Mohini

Plants are tall, height 110-140 cm, flowers at 4-5 node during summer and 5-7 nodes during rainy season after 39-41 days of sowing, fruits five ridges, 11.3-12.6 cm long at marketable stage, suitable for summer and rainy season cultivation; gives yield of 130 -150 q/ha. It tolerate high temperature during summer season and resistant to YVMV under field conditions. This has been identified for the cultivation and release through AICRP for all the okra growing regions of the country.

Kashi Pragati

Plants are tall, height 130-175 cm, with 1-2 effective branches. First flower appears after 36-38 days after sowing on 4th nodes during rainy season and 3rd node during summer season. Fruits are 8-10 cm in length at marketable stage, 23-25 per plants and yield 180-190 q/ha during rainy and 130-140 q/ha during summer season. This is resistant to YVMV and OLCV. It has been released and notified during the XII meeting of Central Sub Committee on Crop Standard Notification and Release of Varieties for Horticultural Crops for the cultivation in Chhattisgarh, Orissa and A.P.

Kashi Satdhari

Plant height is 130-150 cm with 2-3 effective branches, flowering at 42 days after sowing at 3rd to 4th nodes. A plant bears 18-25 fruits with seven ridges, length 13-15 cm at marketable stage and yield 110-140 q/ha, resistant to YVMV under field conditions. This has been notified during the XII meeting of Central Sub Committee on Crop Standard Notification and Release of Varieties for Horticultural Crops for the cultivation in U.P., and Jharkhand.

Shitla Uphar

Plants are medium tall, height 110-130 cm, flowering starts at 38-40 days after sowing at 4-5 nodes. Fruits are green, 11-13 cm long at marketable stage and yield 150-170 q/ha. This is resistant to yellow vein mosaic virus and OLCV. It has been notified and released for the cultivation in Punjab, U.P., Bihar, M.P.

Shitla Jyoti

This hybrid is suitable for warm humid climate with relatively long day length. Plants are medium tall, height 110-150 cm, flowering starts on 30-40 days after sowing at 4-5 nodes. Fruit are green, 12-14 cm long at marketable stage, yield 180-200 q/ha. This is resistant to YVMV and OLCV. This has been released and notified for the cultivation in Rajasthan, Gujarat, Haryana, and Chhattisgarh, Orissa and A.P.

Kashi Mahima

Plants of this hybrid are tall, height 130-170 cm, flowering starts at 36-40 days after sowing at 4-5 nodes, fruits green with 12-14 cm of length at marketable stage and yield 200-220 q/ha. This has shown field resistance against YVMV and OLCV. This has been identified for cultivation and released through AICRP for the cultivation in Punjab, U.P., Bihar, Jharkhand, Chhattisgarh, Orissa, A.P. and M.P.

Kashi Lila

It is early, medium tall (120-140 cm) variety with short internodes along with single or double branch attached in narrow angle with main branch and resistant to yellow vein mosaic virus (YVMV) and okra leaf curl virus (OLCV) under field condition. It takes 30-35 days for first flowering. Fruits are 10-13 cm in length and 9-11 g in weight at marketable stage and dark green in colour. The fruits are available after 35-90 days, total yield being 15-19 t/ha.

Aruna (AE 198)

High yielding (15.8 t/ha) Okra variety KAU, attractive, long, red fruits, average fruit weight is 27 g, average fruit length is 27 cm.

Kiran (AE 1)

High yielding (11.21 t/ha) Okra variety of KAU, attractive, long, light green fruits, average fruit weight is 27 g, average fruit length is 25 cm.

Salkeerthi (AE 202)

High yielding (16.2 t/ha) Okra variety of KAU, attractive, long, light green fruits, average fruit weight is 28 g, average fruit length is 27 cm.

6

Cucurbits-Cucumber

Classification of *cucurbitaceae* family in to different cucurbitaceous vegetable crops:

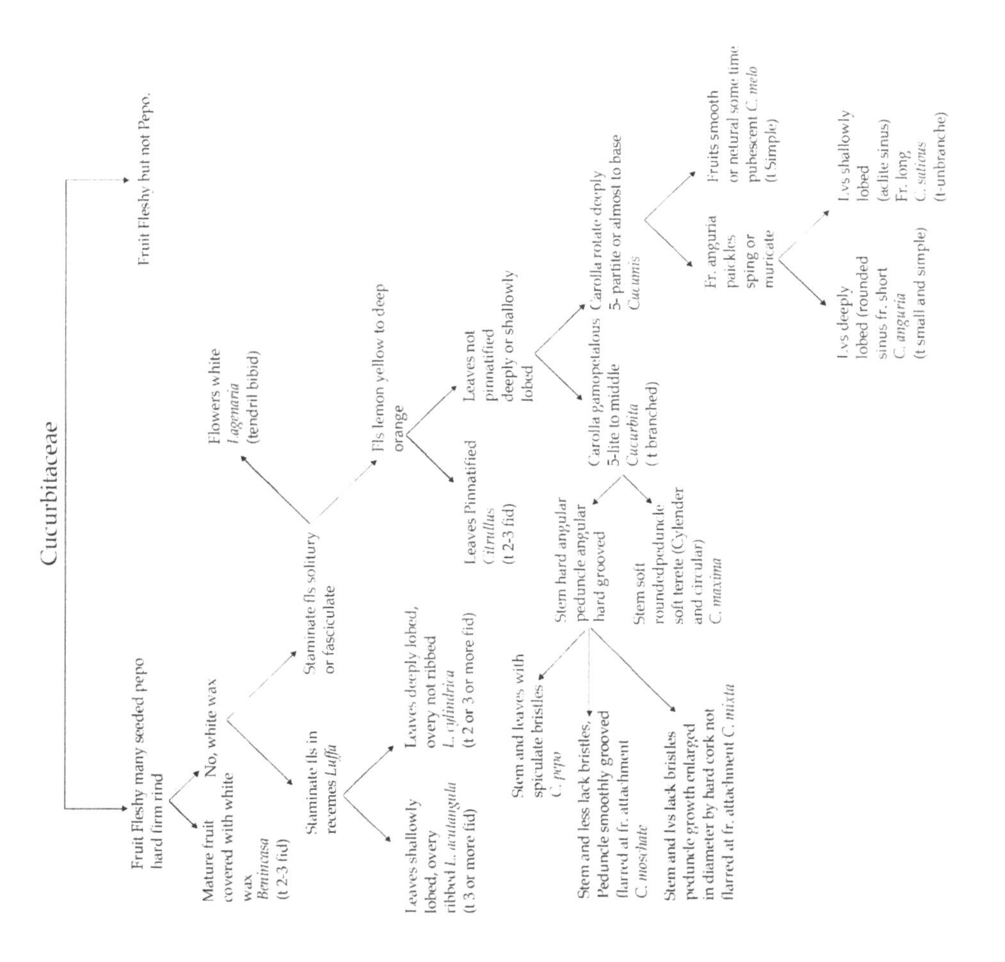

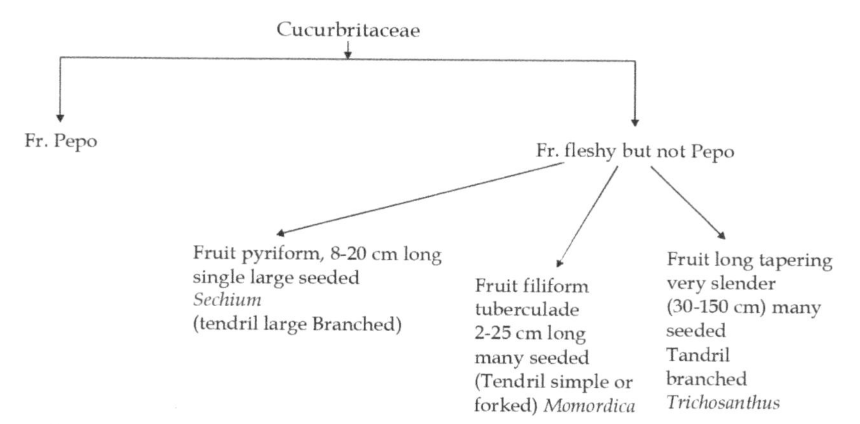

Cucumber

Cucumber (*Cucumis sativus*) is a warm season vegetable used as salad and for pickling purpose has little or no frost tolerance. Growth and development favoured by temperature above 20^{0}C. Cucumber cultivars generally classified as pickling or slicing types. In India often cucumbers (*Cucumis sativus)* are confused with melons (*Cucumis melo*) as morphologically some variants of these species overlap, but they are not cross compatible. Kakri of North India are different from kakri of South India and Maharashtra. In Maharastra and South India cucumbers are referred as kakri but, in North India Kakri means Long melon i.e. *Cucumis melo* Var. *utilissima.*

Major Types of Cucumbers World Over

Pickling type have

1. Smaller length/diameter ratios (*L/D*) than slicers.
2. Lighter green colour skin with more pronounced warts (tubercles) at the immature stage.
3. Most frequently harvested. The *L/D* for pickles is expected to be in the range of 2.8-3.2. As *L/D* is influenced by plant density and plant size breeder has to be cautious. Increase in planting density result in reduction of plant size and lower *L/D.* Small or shorter fruits have lower *L/D.* Big or longer fruit-have higher *L/D.* They should not be confused with gherkins.

Slicers

All slicing cucumbers are called fresh-market cucumber. They are white spined and mostly dark green exterior colour. Most slicers have slightly rounded end and taper slightly for stem to blossom end, although cylindrical shaped fruit with blocky or even rounded end also available. They have length of about 6 to10inch, diameter of 1 1/2 to3 inch, *L/D* is greater than equal to 4.0, The are thick skinned harvested at later stages of maturity.

Parthenocarpic Slicers

They are bred for glasshouse trade, they set and develop fruit without pollination. It is popular in Western Europe. These are 12 inch in length with green coloured skin and marketed at a price 10 times greater than seeded cucumber (similar size and shape). American consumers do not appreciate these because of high cost.

Black spined cucumbers: These cultivars fruit turn yellow-orange or bronze as they increase in size and approach maturation. Commonly they are disliked if slightly over maturated. Resistance to cucumber scab and cucumber mosaic virus (CMV) was made common in Black spined cultivars.

White spine cucumbers: These cultivars fruit turn light yellow on maturation. They are liked even slightly over matured.

Origin and distribution: Believed to be native of India or southern Asia, and has apparently been cultivated there for 3000 years. Burma is regarded as secondary Center of origin and was carried westward to Asia Minor, North Africa, and southern Europe. The progenitor of cucumber, *Cucumis hardwickii* is seen in the foothills of Himalaya. It resembles cucumber except for the smooth fruit surface and extremely bitter flesh. As per De Candolle (1967) cucumber is an indigenous vegetable to India and It is now grown throughout the world in subtropical and tropical climates.

Botany: *Cucumis sativus* L. is commonly a monoecious, annual, trailing or climbing vine (Bailey, 1969). Unlike muskmelon cucumber have hirsute or scabrous stems. It is distinct from of *Cucumis anguria* in having triangular-ovate leaves with shallow and acute sinuses, unlike the deep, rounded lobes of the West Indian Gherkin. Cucumber flowers are usually from 1 to 2 inch in diameter. Staminate flowers are born either singly or in clusters, but pistillate flowers are often solitary in the leaf axes and have shorter and stouter pedicels than staminate flowers. Cucumber fruits are spiny when young and develop into a nearly globular to oblong or cylindrical shape as they mature, turning to a creamy light yellow or deep arrange at seed maturity.

Cucumber has chromosome number $2n = 2x = 14$. Other *Cucumis* species have $2n=2x=24$, $2n=4x=48$, etc. Other related cucumber species is West Indian gherkin, or burr cucumber : *Cucumis anguria* var. *anguria* 2n=24, and it is not cross compatible with *Cucumis sativus. Cucumis anguria* var.longipes has common morphological features of *Cucumis anguria* var. *anguria* (non-bitter) except for bitterness.

Cucumis sativus var. *hardwickii* R.(Alex.) or *Cucumis hardwickii* has chromosome number 2 n=2x =14 and it will cross readily with the cucumber and sets fruits, reciprocal crosses also form in to seeds after fruit set. *Cucumis sativus* var. *hardwickii* is either a feral or progenitor form of cultivated cucumber *Cucumis sativus* L.

The var. *hardwickii* line LJ90430 is being used in many breeding programs. This is larger plant than *Cucumis sativus* cultivars, seems to lack apical dominance and have larger lateral branches. Seed are 1/6th size of seed of *Cucumis sativus* cultivars. Fruits bitter, elliptical in shape and weight 25-to35 g when mature. It is a short day plant produce flowers when photoperiod is less than 12 hours at 30°C day / 20°C Night temperatures.

Interspecific hybridization attempted between Cucumber and muskmelon resulted with germination of pollen, pollen tube transverse length of style, sometimes entering the ovule, but embryos do not develop. (Kho *et al.*, 1980).

Evolution and Domestication

Domestication of cucumber has led to several changes in relation to fruit morphology. Eg. Reduction in fruit ridges number and depth and alterations in shapes. Recent changes accompanying domestication of cucumber is transition from monoecious to gynoecious or predominantly female sex expression for commercial cultivars. Female sex tendency are used to obtain greater uniformity of fruit maturity at harvest and greater early harvest yields.

Different morphotypes in mutants

1. Determinate vine types and compact or bush plant types suited for high density planting and mechanical harvesting.
2. Little leaf a highly branched plant type with small leaves possessing drought adaptations.
3. Glabrous plant types that lack foliar trichomes that are thought to control white fly (*Trialeurodes vaporariorum*) in green house and pickleworm (*Diaphoria nitidalis*) in fields (Pulliam *et al.*, 1979).

World cucumbers are of 4 types

1. Field cucumber with white or black spines.
2. English forcing cucumber – 90 cm long and Parthenocarpic.
3. Sikkim cucumber – reddish brown fruits.
4. Pickling cucumber – conferred with gherkins.

The Bur Gherkin (***Cucumis anguria*** var. ***anguria***, Cucurbitaceae): Known usually as "bur(r) gherkin", "West Indian(n) gherkin", or just "gherkin", *Cucumis anguria* L. ar. *anguria* has been cultivated since "before 1650" (19). Various authors considered the plant to be of New World origin, but Naudin (23), pointing out that all other *Cucumis* species are Old World plants, argued well for an African origin of the bur gherkin and for the introduction of it to the Americas nby Negro slaves (see also 16, 19).

Meeuse (19) concluded that bur gherkin "is a cultigen descended from a non-nbitter variant (mutant) of an African wild species described as *Cucumis longipes* Hook, f., which normally has bitter fruits". He regarded C. anguria and *C. longipes* as conspecific, treating the latter as *C. anguria* var. *longipes* (Hook.f.) Meeuse. The taxa are nearly fully interfertile (8, 19, 26, 29).

Floral Biology and Artificial Selfing and Crossing

Cucumbers exhibit a range of floral morphologies. Staminate, pistillate and hermaphrodite flowers occur in various arrangements, yielding several types of sex expression and are greatly influenced by environment. Multiple staminate flowers on single leaf axils, pistillate flowers frequently occur singularly although occasionally staminate or additional pistillate flowers occur subsequently are some of these variants. A gene conditioning multiple pistillate flowers per node has been reported (Nandgaonkar and Baker, 1981).

Monoecious forms have three phages of sex expression each of variable duration:

I phase-only staminate flowers.

II phase- irregularly alternating female, male or mixed nodes.

III phase-only pistillate nodes.

Pistillate flower are receptive in morning or up to midday on the day they open. Hot dry conditions reduce pistil reception and pollen viability. The success of controlled pollination is enhanced by removal of any previously pollinated female flower. The first fertilized flower inhibits the development of subsequent fruit. In open field one day prior to opening of flower cover with gelatin capsule or tie with wire to prevent them from opening. Next day pollination has to be carried out. Use the entire male flower by removing calyx and corolla and hold it to the stigma of female flower and gently rotate to dislodge the pollen. The female flowers must be covered or closed after pollination to prevent contamination. For this gelatin capsule can be used but it causes abortion due to excessive heat and humidity. Wire tying is ideal. Tagging the female flower completes the process. Tag should last for 60 days.

Genetic studies

1. Spine surface: F/f coarse spine dominant to fine.
2. Spine colour: B/b black dominant to white.
3. Spine number: S/s few dominant to many.
4. Netting on fruit surface: H/h heavy dominant to no netting.
5. Ripe fruit colour: Digenic (R, C) :

Genotype	Phenotype	F2 segregation ratio
R_C_	Red	9
R_cc	orange	3
r C_	yellow	3
rr cc	creamy	1

6. Flesh colour: Digenic (V, W)

Genotype	Phenotype	F2 segregation ratio
V_W_	white	9
V- ww	intense white	3
vv W_	intense yellow	3
vv ww	orange	1

7. Growth habit De/de Indeterminate partially dominant to determinate.
8. Corolla colour Y/y orange dominant to yellow
9. Cluster flowering Cl/cl solitary dominant to cluster
10. Male sterility
    - ms2
    - ms1 fertility dominant to sterility
11. S+ Accelerates femaleness under short day condition
12. I+ Accelerate femaleness under long day condition
13. Ar: resistance to anthracnose dominant.
14. dw: dwarfness recessive.
15. pc (p): Parthenocarpy recessive.
16. ap: apetallous recessive.
17. pm-1: resistance to powdery mildew race 1,
    pm-2: resistance to powdery mildew race 2,
    pm-3: resistance to powdery mildew race 3.

18. dm: resistance to Downey mildew recessive
19. de/de determinate and cp/cp: compact

These reported to have pleiotropic effects on sex expression.

Genetics of Sex Expression

Except *Cucumis sativus* var. *hardwickii* (short day) all cucumbers are day neutral. But short day promote femaleness at early node and also induces frequent appearance of female flowers. Lower temperature increases female tendencies. High temperature and long days promote maleness. Two genes reported for male sterility are ms-1 and ms-2 which condition abortion of staminate flowers.

As per Lower and Edward (1986)

At least 3 major loci involved

a: androecious and m: andromonoecious

1. m+, m m+/- strictly diclinous (unisexuality): Separation of sex i.e. pistillate and staminate flower m/m: result in hermaphrodite flower
2. F+, F: control degree of female tendency

   F/F: Enhances the femaleness and influenced by environment

   F: allele is partially dominant over F+ and intensified female expression
3. a+, a aa: intensified maleness, The effect of this and male intensification is contingents upon F+F+ i.e. ‘a’ is subordinate to ‘F’ locus.

   m+/- Dicliny, m/m-hermaphrodite,

   F+F+: male, a/a: male & a/a androecious.

   ‘F’ is partially dominant over F^+ ‘a’ is recessive & ’m’ is recessive.

When F^+F^+ is there a/a will give only female flower and suppresses the m^+/m gene expression.

Phenotype and genotypes of basic sex types in cucumber (Lower and Edward,1986)**:**

Phenotype	Genotype / locus		
	m	F	a
Androecious	-/-	F^+F^+	a/a
Monoecious	m^+/m^+	F+F+	-/-
Hermaphrodite	m/m	F/F	-/-
Gynoecious	m^+/m^+	F/F	-/-

a is hypostatic to F^+,

Tatlioglu (1993) given detailed account of genetics of sex expression and proposed 5 sex types : Monoecious, Androecious, Gynoecious, Hermaphrodite & Andromonoecious,

Three major genes: Acr/acr, M/m, & A/a

1. Acr – gynoecy, Acr is partially dominant to 'acr'
2. a – androecy & it is hypostatic to gene Acr i.e. it acts only when acr/acr is there,
3. M – diclinous, mm – hermaphrodite.

Acr		acr	
A or a		A	a
M	Gynoecious	Monoecious	Androecious
m	Hermaphrodite	Andromonoecious	Androecious

	Acr	M	a	
1.	Acr/-	M/-	-/-	Gynoecy
2.	Acr/-	m/m	-/-	Hermaphrodite
3.	acr/ acr	M/-	A/-	Monoecious
4.	acr/ acr	m/m	A/-	Andromonoecious
5.	acr/ acr	-/-	a/a	Androecious

To assess this genetics information one need to cross homozygous parents.

Female: Gynoecious Male: Androecious

Acr Acr, MM , AA X acr acr, mm, aa

F1 Acr acr Mm Aa (gynoecious) Selfing

F2 segregation needs 81(64) plants.

Unstable sex expression in gynoecious lines: Homozygous dominant (Acr Acr) is more stable than hetrozygous (Acr Acr).

Gynocious X Hermaphrodite

Acr/Acr, M/M, -/- ↓ Acr /Acr, m/m -/-

F1: Acr/Acr M/m -/- Gynoecious more stable.

Gynoecious X Monoecious

Acr/Acr, M/M, -/- acr acr, M/M, A/A

F1 : Acr/acr, M/M A/- PartiallyDominant Gynocious less stable.

Therefore it is recommended to use Hermophrodite lines for maintenance. That's why go for Gynocious X Hermaphrodite rather than gynocious X monoecious for maintaining gynocious lines.

Unstable sex expression in gynocious lines (Lower and Edward, 1986): Year to year variations within location in female: Male ratio over 3 years study was of a relatively small magnitude when compared to location-to-location variability.

Gynocious (m+/m+, F/F, a+/a+) inbreeds and gynocious X hermaphrodite hybrids (m+/m, F/F, a+/a+) were found to be for more consistent in sex expression across location. Some male flowers were present in all environments but, at a very low frequency (<1%). Development of inbred lines with satisfactory stable sex expression often require more than the transfer of 3 genes. This effects the performance of hybrid in different environments for sex expression. Lines should be assessed at various environments for sex expression and selected for stability across environments.

Chemical Regulation of Sex Expression

The first report in any plant species of alteration of sex expression via exogenous chemicals was in 1949. Auxin application shifted sex expression in cucumber towards femaleness (Laibach and Kribben 1949). Twenty years later, the unsaturated hydrocarbons of ethylene and acetylene were shown to be very potent promoters of female sex expression. Synthetic ethylene releasing compound 2-chloroethylphosphonic acid (ethephon) was found to provide effective means of female flower promotion in both monoecious and andromonoecious cultivars. (Mc Murray and Miller, 1968, Robinson *et al.*, 1969). Ethephon is cheaper chemical.

Male flower promotion

1. GA: Wittwer and Bukovac, 1958: Foliar application of GA3 induced maleness in monoecious cultivars. GA4 and GA7 several times more effective than GA3 in strictly female flowering lines. Also leads to internodal elongation and increasing brittleness of stems.
2. Silver nitrate ($AgNO_3$) is the best as it is less expensive, more effective and more reliable. Silver nitrate with minimum risk of phytotoxicity compared to other chemicals (GA_3), it does not cause inter nodal elongation or brittleness associated with gibberellins.
3. Silver thiosulphate ($Ag(S_2O_3)_2$) is less expensive, more effective and more reliable.
4. Aminoethoxyvinylglycine (AVG)

This has helped breeder to the greater extent for maintenance of gynoecious lines (AgNO3), induction of femaleness in commercial hybrids (ethaphon). Monoecious X monoecious hybrids can be produced without contamination and similarly gynoecious X gynoecious also can be produced.

Gynoecious lines can be planted and monitored for female flower appearance in first 4 to 5 nodes and then treat with chemicals for maintenance. Heterogeneous gynoecious material can be planted and selection for strict gynoecious lines and maintenance can be done easily.

Breeding objectives

1. Increased in yield: sequential fruiting.
2. Earliness.
3. Bush type plants.
4. High female to male sex ratio.
5. Pest resistance.
6. Cold tolerance.
7. Plant stand / Establishment.
8. Improved quality: Low cucurbitacin content, Less seeds, More flesh thickness, Desirable flesh colour.
9. Resistance to diseases.
10. Good transport quality.

Objectives for India: (AICVIP)

- Earliest node producing Pistillate / hermaphrodite flowers i.e. earliness.
- High female to male sex ratio, especially in monoecious types.
- Attractive green or dark green fruits with smooth fruits surface and without prominent spines or prickles.
- Uniform long cylindrical shape without crookneck.
- Fruits free from carpel separation showing hollow spots.
- Fruits free from bitterness.
- Varieties, which do not produce, mature seeds at edible maturity.
- Resistant to unfavorable conditions, pathogen and insects.

Breeding Methods

Introduction: In India introduced varieties are still popular.

eg., Japanese long green: 45 days

Strait: with spines, Poinsette Chinese long.

Mass selection: As the crop is cross pollinated traits like bush habit, spinelessness and free from bitterness can be improved through mass selection.

Bulk population method: Pickling and slicing selections with resistance to anthracnose and downy mildew have been developed from crosses involving anthracnose resistant introduction (PI 197087) and downy mildew resistant lines.

Back cross method: Scab resistant lines have been evolved through back cross breeding. Other characters, which can be exploited are resistance to tobacco mosaic virus in Chinese long (China), mosaic virus resistance in Tokyo Long Green (Japan) and resistance to Downy mildew, Wantee (USA) .

Heterosis breeding: - Heterosis has been exploited for earliness, high yield and quality traits. Hybrids are produced by using gynoecious lines.

F1 seed production: Gynoecious lines 10 rows + male line one row can be planted for hybrid seed production. Gynoecious lines can be maintained by spraying GA 15ppm. Spartan Dawn is the first F1 processing cucumber, developed from gynoecious line MSU71305. Other approaches suggested include use of male sterility or use of selective differential gamaticide, Sodium 2,3-dichloro-isobutarate. Gill et al, 1973 developed F1 hybrid Pusa Sanyog by crossing Kaga Amoga Fushinari (Japanese) with Green Long Naples. Realized 23 to 128 per cent higher yielding and 10 days earlier than Green Long Naples. Tropical gynocious lines were developed by More and Sheshadri (1988).

Polyploid Treeding

Tetraploid cucumber have been obtained by colchicine treatment. They are not of any economic significance mainly due to their low productivity. Triploids also have been produced which are parthenocarpic. Lower fertility and poor yield are major drawbacks of triploids.

Mutation Breeding

Developed male sterile lines by gamma irradiation. Swarna Ageti is developed through mutation breeding (selection in M5 generation).

Interspecific hybridization

Cucumis sativus (2n=14) crosses with other species with 2n=14 producing fertile F1.

Breeding Programs : Cucumber is a crop with narrow genetic base and hence population with broad genetic base would be more responsive.

Outline of S1 progeny selection for fruit yield (Nienhuis, 1982)

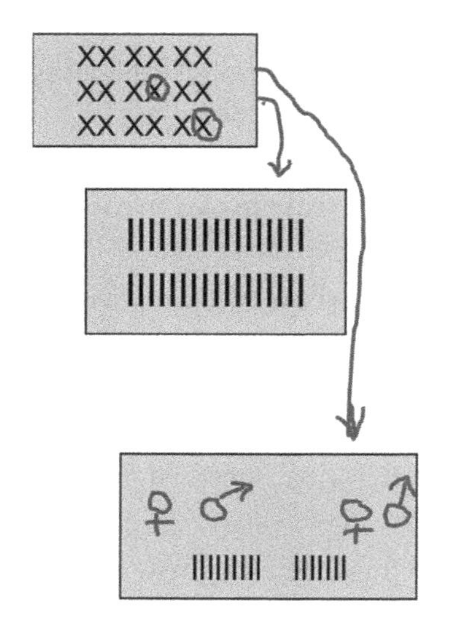

Synthetic population:

Generation of family structure - (Winter, greenhouse)

A) Self pollination 100 individuals in each population

Progeny test - (Early, summer, field)

A) Progeny test S1 families for fruit yield

B) Select top 15 percent

Recombination block - (Late summer, field)

A) Plant approx 2-3 weeks after progeny test planning

B) Each family represented in a Random male and female row

C) Before flowering, rogue out Non selected families

D) Harvest seed from only Female rows

Outline for recurrent selection for specific combining ability for fruit yield in a cucumber population

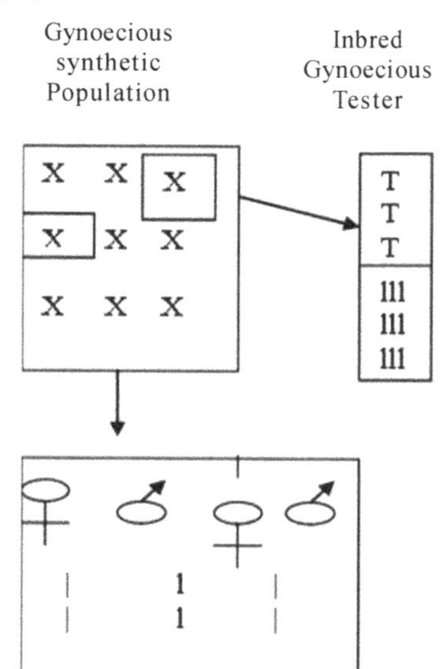

Generation of family structure - (Winter, greenhouse)

A) Self and produce testcrosses Between 150 individuals with Gynoecious Inbred line. ***GY14***

As a tester

Progeny test - (Early, summer, field)

A) Evaluate testcrosses Progeny for fruit yield

B) Select top 15 percent

Recombination block - (Late summer, field)

A) Plant approx 2-3 weeks after progeny test planning

B) Each family represented in a Random male and female row

C) Before flowering, rogue out Non selected families

D) Harvest seed from only Female rows

1. *Cucumis trigonus*: 2n=14 : source of resistance to *Dacus cucurbitae* (Fruit fly).
2. *Cucumis sativus* var. *hardwickii*: - 2n=14 for high yield and good adaptability

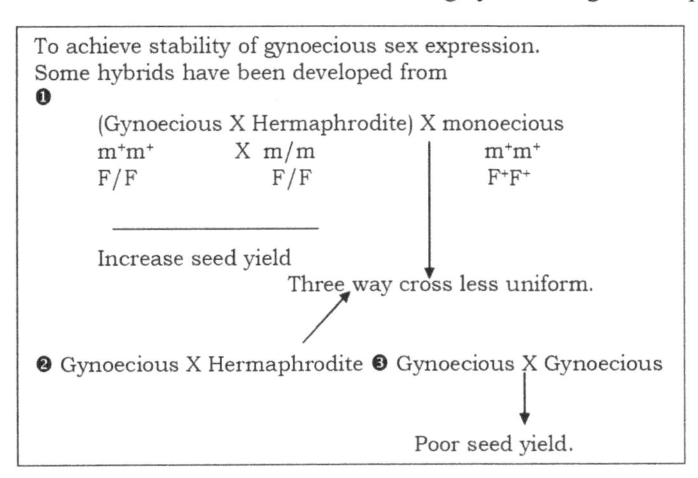

Other species difficult to cross with cucumber are *Cucumis metuliferum* (2n=24) African horned cucumber, *Cucumis dispaceus:* -(teasel gourd) (2n=24) and *Cucumis hirsutum (*Perennial and diocious).

Warmer climates result in more rapid fruit growth and demand a higher level of fruit quality such as fruit firmness, seed cavity size and maturation rate and external colour. It requires large sample size because within highly inbred lines fruit to fruit variation can be seen.

Population Improvement for Inbred Extraction

Cucumber has narrow genetic base and hence population improvement approaches invited. Rec. selection for yield for several cycles keeping variability intact on other traits. For this random mating or strategic crosses (Pedigree breeding) is must otherwise so long term and medium term goals achieved simultaneously. Before extracting inbreds, it is important to ensure that the frequency of genes for other desirable traits is great enough to recover them at an early inbreeding stage. Select lines for derived character at segregation stage and then advance to inbred extraction. Evaluation and selection during inbreeding is important to achieve maximum gains. As inbreeding progresses, plot sizes must increase to allow discrimination of suitable fruit quality difference. Replication across environment should also be expanded to evaluate adaptability and stability of lines.

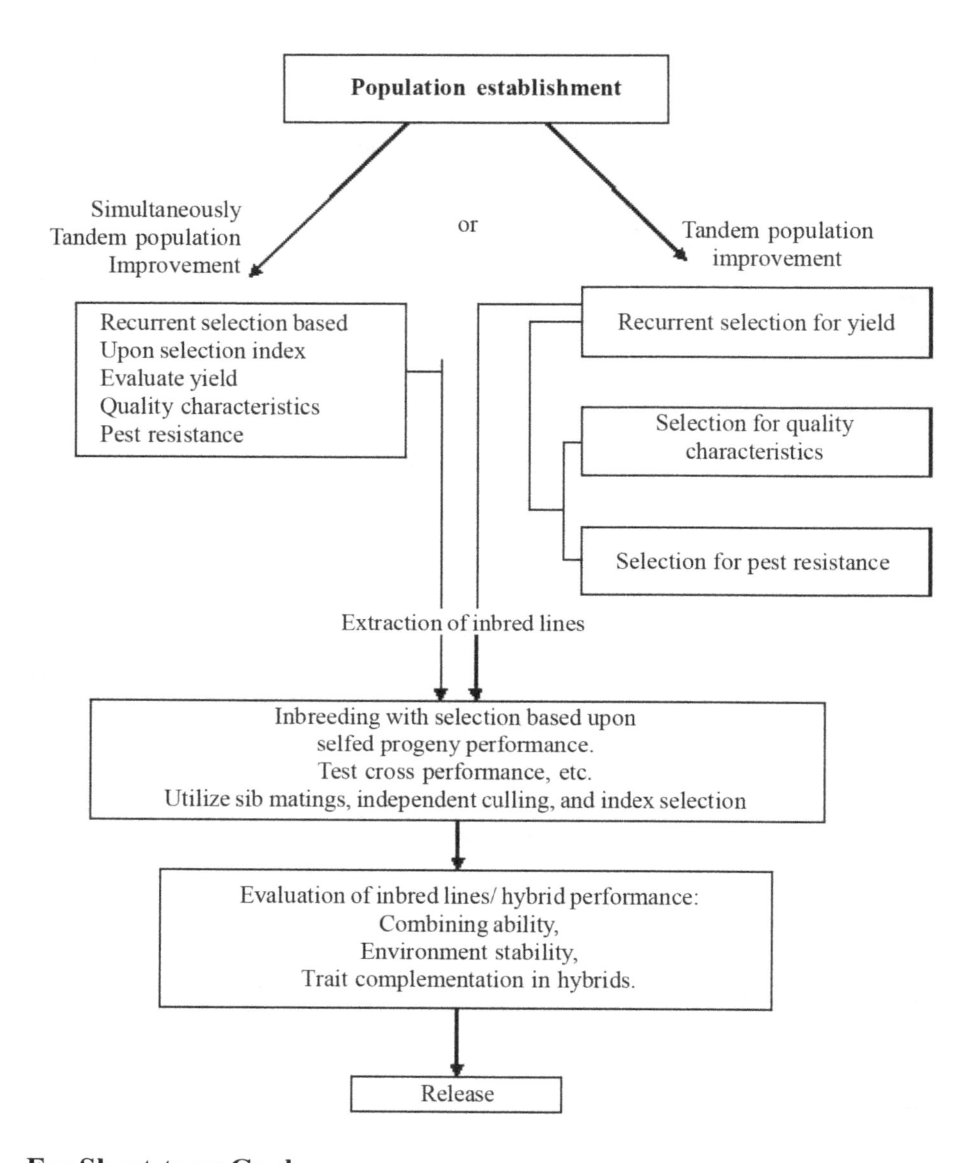

For Short-term Goals

If short-term gains are also sought of breeding scheme (back cross, mass selection etc.) with narrow germplasm base may be conducted simultaneously. Development of inbreds from F2, then go for controlled pollination. Screen as many as 10 F2 population to develop 10 to 25 F3 population for each. Eliminate 50 per cent of plots based on first harvest fruit character and at second harvest totally 75 per cent eliminated. Continue through F4 and F5 generation with increasing pressure on inbred material for quality character, stable sex expression, seedling vigor, etc. For identification of inbred lines F_5 and F_6 lines can be used for test crossing to test gca and sca. Compare the hybrids in different environment. It can be attempted for different dates of sowing in same season to evaluate genotypes for stability. Crossing of desirable lines to a recurrent parent (back cross) or a new parent for a positive genetic contribution simply delays the testing stage but, is an important aspect of inbred or parental development.

Breeding for Specific Objectives

Herbicide tolerant cultivars: Some vine crops do posses limited resistance to triazine herbicide compound. There is good scope and it is more exciting and potential for development of allelopathic cultivars that essentially contain built in mechanism for weed control.

Improving fruit quality: Skin colour, spine colour, presence or absence of warts and spines on fruits are important quality traits which can be considered for improvement.

Fruit shape: Related to cosmetic needs or desires of various segments of industry.

Flesh firmness is a positive attribute, as is skin (exocarp) tenderness.

If skin is very tough then it is unpalatable in either pickling or slicing products but offers little protection against the rough handling at harvesting, sorting, grading and shipping. Tender skin is desirable as edible product but cannot resist rough handling and also is conductive to weight loss (H_2O evaporation).

Fruit firmness, related to both flesh firmness and seed cavity (endocarp) or locule size. Higher locule size lower fruit firmness. Slender fruit (large L/D) has less endocarp (seed cavity). Lower L/D ratio associated with higher endocarp area. There is positive and linear relationship between rapid fruit maturation and seed cavity growth. These two characters are negatively correlated with fruit firmness and vine storage time for marketable fruits. Increase in fruit firmness and decrease in seed cavity size are frequently accompanied by placental hollowness. This is unattractive and undesirable lead to processing problems. Carpel separation is associated with rapid fruit enlargement and less firmness in cultivars of large sized fruits. Increases in flesh firmness tend to reduce occurrence of carpel separation.

Fruit firmness: Magness Taylor Fruit Pressure Tester (exo-endo and mesocarp all enclusive) used to assess fruit firmness.

Fruit quality

1. Ratio of cavity diameter to fruit diameter.
2. Incidence of placental hallowness.

3. Incidence of carpel separation.
4. Off types occurance: 2 or 4 carpelles.
5. The rate of seed maturation.

Stable Gynocious lines: Desirable for commercial cultivation. Determinate (de) and compact plant (cp) genes which are known to reduce plant size also result in reduced male or staminate expression in gynoecious hybrids. This enhance femaleness may translate into increased gynoecious stability and higher fruit numbers.

Multiple disease resistances: Anthracnose, Downy mildew, Scab, CMV, Leaf spot, Powdery mildew, Bacterial wilt, WMV-142 (Water melon mosaic virus), Cucumber green mottle virus. Fruit rot caused by *Rhizoctonia, Pythium and Phomopsis* where high level of resistance is lacking. Resistance to nematodes is totally lacking and resistance is also lacking for *Fusarium* wilt.

Achievements

China: Medium late, straight variety, very hardy and prolific. Fruit very long (50 cm), slender, with deep green skin, white spined, flesh is white, firm and crisp.

Local cultivars: Poona khira - Poona regions.

Balam Khira - UP (Saharanpur area).

Kakri - Maharastra area

Belgaum local, Greenlong, Whitelong, Sheetal.

Exotic cultivars: Moviri - Moscow - Hybrid.

Redlands long white - Queensland, Australia.

Ventura - Netherlands.

Parthenocarpic varieties - Paragon, Parthenon

Pickling types: Vesta, Balam Khira

Cucumber (Swarna Ageti)

- Developed through mutation breeding and selection in M-5 generation
- Fruit: Cylindrical long (150-200 g), green and without prominent placental hollowness
- Tolerant to powdery mildew
- Time of sowing: July-August and February-March
- Spacing: 2m x 0.5 m
- Maturity : First harvest 45-50 days after sowing
- Yield : 300-325 q/ha
- Recommended for Bihar, Jharkhand, Uttar Pradesh and Punjab

Swarna Sheetal

- Developed through hybridization (Hot season x Long Green) followed by pedigree and recurrent selection
- Fruit: Cylindrical long (200-250 g), greenish white and without prominent placental hollowness
- Tolerant to powdery mildew
- Time of sowing: February-March
- Spacing: 2m x 0.5 m
- Seeds rate: 600-700 g/ha
- Maturity : First harvest 60-65 days after sowing
- Yield : 300-325 q/ha
- Recommended for Jharkhand, and Bihar

Swarna Poorna

- Developed through pure line selection from CH-20
- Fruit: Cylindrical long (300 g), light green and without prominent placental hollowness
- Tolerant to powdery mildew
- Time of sowing: July-August and February-March
- Spacing: 2m x 0.5 m
- Seeds rate: 600-700 g/ha
- Maturity : First harvest 55-60 days after sowing
- Yield : 300-350 q/ha
- Recommended for Bihar, Jharkhand, Uttar Pradesh and Punjab

7

Musk Melon

Melons (*Cucumis melo*) are annuals with climbing, creeping or trailing vines of length up to three meters. The desert types quench thirst and add to the nutrient content of main diet. The non-desert types are used as vegetables. Muskmelon (*Cucumis melo*., 2n=2x=24) encompasses the netted, salmon -flesh cantaloupe, the smooth – skinned green fleshed 'Honey Dew', the wrinkled, white – fleshed, 'Golden Beauty' and several other dessert melons in USA.

Origin: Not known with certainty but, 40 or more wild species of *Cucumis* are found to occur in the tropics and subtropics of Africa. Wild species of *Cucumis* occur in Africa and it is likely that it originated in African continent. It does not appear to have been known to the ancient Egyptians, not to the Greeks and seemed to have reached Europe towards the decline of Roman Empire. Although melon was introduced into Asia at a comparatively later date there are undoubtedly well developed secondary centers of origin of *C. melo* in India, China and then Southern USSR. In India Oriental pickling melon (*C. melo* var *conomon*) and snap melon (*C. melo* var *momordica*) are unique and have considerable variability in Western Ghats of India. They are popular desert and preserving melons cultivated in Warm humid tropical condition of South India. It is also considered that China as primary center of origin and Indo-Burma as secondary center of origin. It has been introduced to Europe and from Europe to the USA by the early travelers. At present muskmelon is being cultivated throughout the world under tropical and subtropical climatic conditions. Polymorphism in leaf, flower, fruit shape and colour (Kirkbride, 1993), allowed the classification of horticulturally important melons into seven groups.

Important Melon Groups Grown in Different Countries

1. *C. melo* var. *cantaloupensis* Naud: medium size fruits, round shape, smooth surface marked ribs, orange flesh, aromatic flavour and sweet.
2. *C. melo* var *reticulates* Ser.: medium size fruits, netted surface, few marked ribs, flesh colour from green to red orange.
3. *C. melo* var. *saccharinus* Naud.: medium size fruits, round or oblong shape, smooth lacking the typical musky flavour. These fruits are usually late in maturity and longer keeping than cantaloupensis.

4. *C.melo* var. *flexuosus* Naud.: long and slender fruit eaten immature as an alternative to cucumber.
5. *C.melo* var. *conomon* Mak.: small fruits, smooth surface, white flesh. These melons ripen rapidly, develop high sugar content but little aroma.
6. *C. melo* var. *dudaim* Naud.: small fruits, yellow rind with red streak, white to pink flesh.

Possible evolution of muskmelon

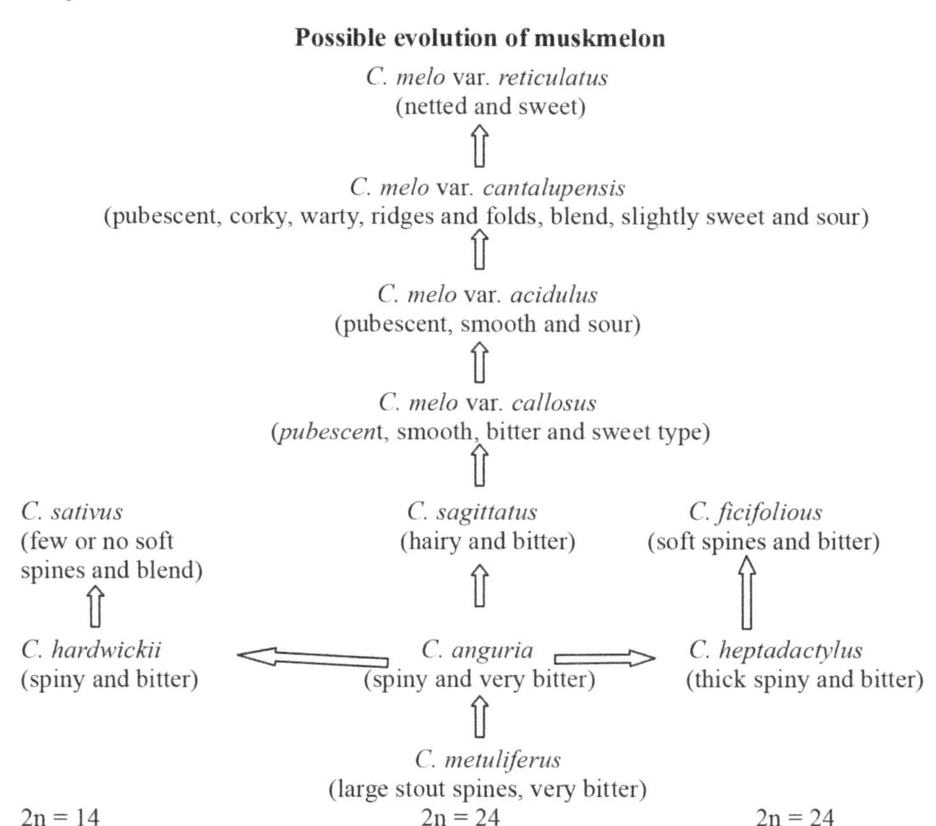

Classification of Variability

1. Cantaloupe melon of Europe: thick, scaly, rough, rind, often deeply grooved.
2. Muskmelon grown in USA: Smaller fruits and rinds which are finely netted to nearly smooth and very shallow ribs.
3. Cassaba or winter melon: This produces large fruits which mature late and have good storage quality. Rind is usually small often stripped or splashed and yellow in colour. Flesh is firm with little musky odour or flavour. Eg. Honey Dew of USA.

4. Oriental pickling melons: Often with elongated fruits resembling cucumber. In India and East used as vegetable.

Based on fruit taste

1. *C. melo* var. *Callosus* – bitter 2 *C. melo* var. *acidulus* – sour 3 *C. melo* var. *cantalupensis* – blend 4 *C. melo* var. *reticulatus* - sweet

Growing regions

1. *C. melo* ssp. *rigidus* – hot dry region (Central Asia and Iran)
2. *C. melo* ssp. *orientale* – hot and humid (Asia Minor)
3. *C. melo* ssp. *europees* – cool and dry region (Asia Minor)
4. *C. melo* ssp. *flexuosus* – semi feral melon – snake melon
5. *C. melo* ssp. *chinensis* – Chinese melon (East Asia)
6. *C. melo* ssp. *spontaneum* – wild melons

Russians assigned Separate genus status for musk melons.

Like: *Melo* with 3 sections: 1) Eumelo 2) Melonosids 3) Bubalion

Indian cultivar of musk melon:

Non dessert types: Phoot (Snap melon)

SI : It is desert type Kachri / Kachauri : Cooked in Nortn India

Dosakaya and Budamkaya in Andra Pradesh, cooking types are called cucumber

Botanical Varieties of Musk Melon

It includes both dessert as well as cooking and salad types used like cucumber.

C. melo Var. *agrestis*: inedible fruit

Var. *Cultura*: Stouter vines

Var. *reticulatus* : netted melons

Var. *Cantaloupensis:* Hard skin, wart surface, netting absent

Var. *indorus* : Winter melons of USA Comprises Honey dew (Green flesh), casaba and *Cremson*, musky odour, smooth fruit surface,

Var. *flexuosus*: long melon (snake or serpantine melon) (Var. *utilissimus*) kakri of NI: like salad cucumber

Var. *Canomon*: Pickling melon, smooth fruit, various shapes.

Var. *Chito*: lemon cucumber, small size referred as vegetable orange

Var. *dudain*: Chito like fruits, fragrant fruits, ornamental type, rich brown fruit surface

Var. *Momordica* Snap melon (Phoot of NE India)

AICRP on vegetable crops : Multilocation centres: Akola, Ludhiana, Hisar, Modipuram, Anand, IARI, Delhi, Faizabad and Durgapura.

Genetic Resources: Durgapura, Faizabad and Ludhiana centers have been give responsiblity for collection and evaluation of germplasm of Muskmelon.

Cytogenetics: Studies on chromosome morphology and meiotic behaviour in muskmelon cultivar of both desert and nondesert nature have indicated that the chromosome number was stable. 2n= 24 (Ramchandra et al, 1985). Desert types showed higher chiasma frequency than the non desert types. The same workers have also demonstrated chromosomal variation in muskmelon on the basis of their experiments with a bushy var Bokor. Tetraploids 2n= 48 induced by colchicine were studied cytologically (Batra, 1952) and compared with the original diploids where fruit quality was superior in tetraploids. Cantaloupe is a polyploidy with 2n=48.

Genetics

Sex expression

From the crosses of hermaphrodite X monoecious types, it has been usually concluded that F2 segregation conforms to a typical digenic ratio as follows :

G – A - = + + = 9 monoecious

G – aa = + a = 3 andromonoecious

gg A- = g + = 3 gynomonoecious

gg aa = g a = 1 hermaphrodite

However, it should be kept in mind that environmental factors and interaction with other genes may give rise to various other sex forms in muskmelon. Usually, there is an association between fruit shape and sex form. Fruit shape of monoecious and gynoeceious lines is oblong and perfect flowers produce round fruits, however, exceptions have also been reported (McCreight *et al.* 1993).

Heteromorphic chromosomes also identified which are responsible for male and females. Female: xx, male: xy.

Gene symbol	Character	Geners
dc1	Resistance to *Dacus cucurbitae*	Single recessive gene
Gf	Green flesh colour	Single dominent gene
Gp	Green petals	- do -
Jf	Juicy flesh	- do -
ms1	Male sterile 1	Single recessive gene
R	Red stem	single dominant gene
O	Oval fruit	- do -

Bi	Bitterness	- do -
Gl	Glabrous	Single dominant gene
N	Nectar less	- do -
So	Sour	- do -
St	Striped pericarp	- do -
Wi	White fruits (immature)	- do -
Y	Yellow pericarp	- do -
W	White fruits (Immature)	- do -
Ag	Tolerant to aphids	- do -
Gp	Green petals	- do -

Rind colour: Yellow monogenic and partially dominant over cream (Bains, 1960: Sandhu, 1990). Dark green dominant over green, Greenish grey dominant over light green, Stripped and mottled pericarp dominant over white (Dutta, 1967), Green monogenically dominant over yellow (Chadha *et al.*, 1972).

Flesh colour: Orrange monogenically dominant over light green (Sandhu, 1990), White monogenically dominant over green (Chadha *et al.*, 1972), Maternal effects (Ramswamy *et al.*, 1977). gf: green flesh, jf: juice flesh.

Flesh thickness: Thinner partially dominant over thicker and monogenically (Bains, 1960).

Flesh taste: Monogenic with partial dominance (Sandhu, 1990).

Juiciness: Less juiciness monogenioc dominant over juiciness (Chadha *et al.*, 1972).

Fruit slipping: Slipping monogenically dominant over non slipping from the stalk (Sandhu, 1990).

Rind craking: Digenic witrh complementry and suplementry action (Sandhu, 1990).

Ribbing: Prounounced and regular ribbing dominant overnon ribbing (Dutta, 1967).

Sutures: Sutureless monogenically dominant over sutured (Sandhu, 1990).

Absence of sutures digenically dominant with supplementary relationship (Chadha *et al.*, 1972).

Netting: Digenic with netting dominant over smoothness. N: Netting, n: Smoothness.

Netting monogenically and partially dominant over smoothyness(Sandhu, 1990),

SS: Supress netting (Ramswamy *et al.*, 1977)

Resistant to fruit fly: Susceptability digenic with suplementry gene action (Sambandam and Chelliah, 1972). F2: 9 susceptible 7: resistant.

Resistance to red pumpkin beetle: Resistant monogenically dominant over susceptibility (Vashista and choudhary, 1974).

Resistance to Powdery mildew: Resistant dominant over susceptibility (Nath *et al.,* 1973).

Resistant recessive to susceptibility (Choudhary and Sivakami, 1972).

Quantitative characters: Earliness: Dominant variance more than the additive varience, over dominance also present (Chadha *et al.*, 1972).

TSS: Partial dominant accompanied with adequate additive genetic variance (Chadha *et al.*, 1972).

Number of fruits/ vine: Partial dominant along with adequate additive genetic variance (Chadha *et al.*, 1972).

Fruit weight: 39 genes: Small fruited partially dominant over large fruited ness (Sambandam and Chelliah, 1972).

Total yield

1. Dominant variance more than additive variance, over-dominance also present (Chadha *et al.*, 1972)
2. Dominance (Kallo and dixit, 1989)
3. Additive gene effects (Sandhu, 1990)

Vine length: Both additive and dominant genetic variance significant.

Dominant variance higher than additive variance (Singh *et al.*, 1989).

Genetic Resources

C. callosus: Resistance source for fruit fly (Chellaih, 1970) and cross compatible with cultivated varieties of musk melon and Resistant to *Fusarium* wilt: (Sambandam and Chelliah, 1972)

Cv: Casaba showed high degree of tolerance to red pumpkin beetle (Vashista and Choudhary, 1972)

Pusa Sharbati moderately resistant to powdery mildew.

Campo, Jacumba, Lerlita, PMR-5 and PMR-6 are resistant sources for powdery mildew (Choudhary and Sivakami, 1972)

Good source of resistance for downy mildew: Buduma Type-1,-2 and -3, Phoontee, Goomuk, Nakkadosa, Ex-2 etc.

According to McCreight *et al.* (1993) and Dhiman (1995) the sources of resistance to disease and insect-pests in muskmelon are given below.

Disease	Source
Powdery mildew	PMR 45, PMR 450, PMR 5, PMR 6, PI
Downy mildew	124111, Seminole
Fusarium wilt	MR 1, PI 414723
CMV	MR 1, PI 124111 F, CM 17187
WMV	PI 161375 PI 414723

Disease/ Insect	Source
Cucumber Green Mottle Mosaic Virus	VRM 5-10, 29-1, 32-1, 42-4, 43-6
Fruit fly	*C. callosus*
Melon aphid	PI 161375, PI 371795
Multiple disease	MR 12.
Resistance (downy mildew, powdery mildew, muskmelon mosaic)	

Objectives: Major breeding objectives are

- Varieties with high total marketable yield and earliness.
- Varieties with thick skin, thick flesh and good consistency.
- Varieties with good flavour, attractive outer colour and flesh colour with good texture.
- Varieties with attractive fruit shape having small seed cavity, small and negligible hollowness.
- Varieties with sweet, juicy and flavorsome fruits are preferred: TSS >10 % (Flesh colour and texture)
- Resistance to insect pests: Red Pumpkin Beetle, fruit fly and Aphids.
- Resistance to powdery mildew and downey mildew, CMV, squash virus, cucumber green mottle virus.
- Hybrids with high yield and good quality.

Breeding Methods

Mass selection : Arka Jeet, Arka Rajhans and MH1 were developed by single plant selection and later maintained by mass selection.

Pedigree method: Pusa Sharbati developed by pedigree method of cross Kuttana x American Cantaloupe

Punjab Sunehri is developed through pedigree method of cross Hara Madhu x Edisto

Other suggested methods are bulk population method and backcross method.

Heterosis breeding : Heterosis is observed for days to fruit harvest, early yield, soluble solid content, sugar content, fruit weight, keeping quality, transportability, fruit flavor, number of nodes, vine length and total yield.

F1 hybrids are produced using monoecious lines as maternal lines. Male sterility also has been exploited. The gene ms1 was discovered in cross Georgia-47 x Smith Perfection. Hybrid seed set is low (15-20%). Nandapuri *et al.*, 1974 found that Arka Rajhans and Har Madhu among the males and MS1 and MS2 among the females were good combiners for maximum number of economic characters. Pandey and Kaloo (1976) reported monoecious hybrids. The highest amount of heterosis (433.72%) over the better parent for early yield was reported by Mishra and sheshadri,1985. Dixit and Kaloo (1983) observed heterosis over better parent for number of fruits per plant to the extent of 54.3% in Punjab Sunehri X Sel 1 and for fruit yield 46.70% in Pusa sharbati X Sardar Melon and 38.1% in Pusa sharbati X Punjab Sunehri.

F1 hybrids are produced by using monoecious lines as maternal lines male sterility also has been exploited. Hybrid seed set is low (15-20%). Measures to improve seed set are only one or two growing main shoots may be retained. Only two or four female flowers on a plant be hybridized and after pollination growing tips removed. Breeding nursery may be kept at 26-30°C. Eg. Punjab Hybrid (F_1): MS1 X Haramadhu: 2 : 1 row ratio for seed production.

Interspecific Hybridization

The botanical varieties of *C. melo* are inter crossable. Attempt made to cross with cucumber has failed. Somatic hybridization suggested. Mentor pollination, growth regulator (benzyl adenine) and use of n-hexane can help in over coming cross incompatibility.

Breeding Achievements

Local cultivars: Kanpuria, Jogia, Mathuria, Batti, Kajra, Jaunpuri, Mahaban and Lucknow safeda in UP. Baghpat melon of Meerut, Jaipuri Netted and Mau melon of Azamgarh in UP. Tonk melon, Sanganer, Haragola and Motea in Rajasthan. Kharri melon of Hoshangabad. Jalgaon, Kabri, Gurbeli and Kavita in MP. Goose melon, Jam or Neel in Maharastra. Ladoo, Papaya, 'Sharbat-e-Anar, Chiranji and Bathesa in Andhra Pradesh. Sankheda and Balteshwar in Gujarat. Kadapa in Karnool. Kutana and Bhagpat in Haryana. Har Dhari and Musa in Punjab.

Exotic cultivars: Allrora - USA and Poongmi – Korea.

Resistant varieties

- Powdery mildew: Campo, Jacumba
- Downy mildew: Georgia 47, Gulf Stream
- Watermelon Virus 1 : B665
- Muskmelon mosaic: Oriental pickling melon

High Yielding Varieties

Arka Jeet : Improved on local collection from Lucknow (IIHR 103).

Arka Rajahans : PLS from local collection from Rajastan (IIHR107).

Hara Madhu, Pusa Sharbati, Pusa Madhuras, Durgapura Madhu, Punjab Suneheri, NDM1 and NDM2, MHY-3 (Durgapura)

Arka Rajhans: (IIHR) It is a midseason variety bearing large oval fruits weighing above 1 kg with fine netted creamy white skin. The flesh is white and sweet and fruit has transportable quality, tolerant to PM. IIHR for southern region. Selection from a local collection (IIHR-107) from Rajasthan. TSS 11-14%. Early maturing, yields 285 q/ha.

Arka Jeet: (IIHR). A selection from Bati strain of UP. Animprovement over a local collection (IIHR-103) from Lucknow. The fruits are flat, small weighing 300-500g orange to orange brown skin, white flesh, big seed cavity, very sweet 15-17% TSS. 150 q/ha in 90 days. Rich in Vit C (41 mg). An early cultivar very similar to Lucknow safed (UP).

Pusa Sharabati: (IARI) An early cultivar maturing in 85 days, with round fruits having netted skin. Salmon-orange flesh is firm and thick with small seed cavity, moderate sweet 11-12% TSS. developed from the cross between Kutana (Haryana) and Resistant No.6 of the USA. 150 q/ha. Suitable for river bed cultivation and other field conditions in northern India.

Pusa Madhuras (No.445): (IARI) It is a midseason selection from a Rajasthan collections with roundish flat fruits weighing a kg or slightly more. The skin is pale green, sparsely netted with dark green stripes and salmon orange flesh, juicy and sweet (12-14% TSS). Keeping quality poor. 150 q/ha in 90-95 days.

Hara Madhu: (PAU) A late cultivar from a local collection of Haryana. The fruit is globose with dark green stripes, weighing one kg, flesh is light green, juicy, very sweet (with 12-15% 1'SS). Keeping quality poor. Fruits round, tapering toward the stalk end, with 10 prominent green sutures. It is susceptible to PM and DM. 125 q/ha.

Punjab Sunehri: (P AU) Is a selection from the cross Hara Madhu x Edisto, early maturing, pale green, thick skin, flesh salmon orange, thick with moderate sweetness 11-12% TSS. Excellent for transport and qualities. It is tolerant to PM and DM. 160 q/ha.

Punjab Hybrid: (PAU) This is an F1 hybrid between a male sterile line (MS1) and Hara Madhu. An early maturing hybrid with orange flesh and netted skin. 12% TSS. Storage and transport is good. Moderately resistant to PM and fruit fly. 160 q/ha.

Durgapur Madhu: A very early cultivar confined to Jaipur region of Rajasthan. The fruits are oblong, weighing 500-600 g, pale green rind, light green flesh with dry texture, very sweet 13-14% TSS, seed cavity big. Recommended by Dept. of Agri. Jaipur, Rajasthan

Swarna (Cantaloupe): Hybrid from IAHS. Fruit is yellowish orange in colour with very sweet, dark orange flesh inside. It can withstand long distance transport.

Sona (Cantaloupe): Hybrid from IAHS. Fruits are closely netted, slightly ribbed and orange cream coloured. Tolerant to PM and DM and possesses good keeping quality.

Mudicode (CS 26): High Yield (30 t/ha). Large, attractive golden yellow fruits, Average fruit weight : 2.15 kg, Average fruit length : 31 cm, High yielding Oriental pickling melon variety KAU.

Saubhagya (CM 8): High Yielding(17.1 t/ha) of KAU, Small to medium sized, oblong, golden yellow fruits, Average fruit weight : 1.1 kg, Average fruit length : 20.5 cm.

Arunima (CS 1): High Yield (27 t/ha, Large, uniformly cylindrical, golden yellow fruits, Average fruit weight : 2.3 kg, Average fruit length : 33.14 cm, High yielding Oriental pickling melon variety of KAU.

Kashi Madhu: Medium vine and leaves sparsely lobed and dark green, fruits are round, with open prominent green sutures, weight 650-725 g, half slip in nature, thin rind, smooth and pale yellow at maturity, flesh salmon orange (mango colour), thick, very juicy, T.S.S. 13-14%, **Long storage with good transportability, tolerant to powdery and downy mildew,** medium maturity and yield 200-270 q/ha, Recommended for the cultivation in U.P., Punjab, and Jharkhand.

Snap Melon (Phoot)

1. **Pusa Shandar:** Developed at IARI, New Delhi, Suitable for sowing in summer and kharif seasons in the plains. This variety can be grown as alone or in intercropping, it es early variety. Fruits are cream coloured, oblong, flesh in light pink. fruits weigh 700g and yields 30 to 38 t/ha and is suitable for growing in home and kitchen gardens.

2. **Grism Bahar:** Suitable for sowing in summer season, fruits are 30-40 cm long and thick with green colour which turn to yellow on ripening.
3. **Kwari Bahar:** Suitable for sowing in kharif season, fruits 25 to 30 cm long and thick with green stripes and turn to yellow on ripening.

Long Melon

1. **Pant Kakri-1:** It is developed at GBPUAT, Vegetable Research Centre, Pantnagar. Selection in the indigenous germplasm released during 2001. Suited for February to April sowing. Fruit are long, light green and straight. Yields 30 t/ha and is free from common diseases and insect-pests. It takes 90 days for seed to seed stage.
2. **Karnal Selection-1:** It is developed at HAU, Hisar, Haryana and is Prolific bearer. Fruit are tender, light green, long, thin, flesh crisp with good flavour and yields to 10 t/ha.
3. **Punjab Long melon-1:** It is developed at PAU, Ludhiana during 1995, fruits are long, thin and light green in colour and yields 86 t/ha.
4. **Arka Sheetal:** It is developed at IIHR, Bangalore. It is a pure line selection from local material (IIHR 3- 1-1-1-5-1) collected from Lucknow and released during 1984. Fruits are light green with medium long and weigh 80 to 90 g. It yields 35 t/ha and fruits are free from bitter principles.

Round Melon or Indian Squash

1. **Arka Tinda:** It is developed at IIFIR, Bangalore. It is an advanced pedigree selection of the cross between T3 (from Rajsthan) and T8 (from Punjab) and released during 1984. It is early maturing & suited for summer cultivation. It is ready for harvesting after 5 to 6 days after pollination. Fruits are light green, round soft hairs present on fruits. It yields 10 t/ha.
2. **White Green Tinda:** It matures in 50 to 55 deays after sowing and fruits are medium sized with light green colour and yields 20 to 25 t/ha.
3. **Dark Green Tinda:** It bears medium sized fruits with dark green skin and yields 20 to 22.5 t/ha.
4. **Tinda S-15:** It matures after 65 to 70 days of sowing and bears medium sized fruits with light green skin and fruits are flatish round and weight 50 g. Flesh colour is white and yields 25 t/ha.
5. **S-48:** It is a selection from local material & matures in 60 days after sowing. Fruits medium sized, with light green skin, flattish round shape, shiny and

pubiscent and weight 50g. Flesh colour is white and bears 8 to 10 fruits/plant and yields 45 t/ha.

Other varieties: Punjab Tinda, Tinda Ludhiana, Chhoti Kakri, Badi Kakri, Lucknow Early, Laila Ki Angulia, Majnu ki Pasalia.

8

Water Melon

Water melon is tender to frost and most of cultivars require relatively long growing season. Plant growth and fruit development are fallowed by higher temperature and abundant sunlight. Atmospheric humidity greatly regulates fruit and foliage diseases.

Taxonomy and Origin: Ealier it was classified as *Citrullus Vulgaris* Schrod. But in 1963 Thieret called attention to the correct name, *C. lanatus* (Thumb) Mansf. Africa is centre of origin of the genus. In 1857 David Living Stone, the famous missionary explorer found both bitter and sweet melons grown together in Africa and native people used them as source of water. Cultivation is prehistoric as there are pictures made in ancient Egypt. Early cultivation occurred in Mediterranean area and as far as India also and It was brought to USA by European colonist as early as 1629.

Species descriptions and cross compatibility

Cognìaux and Harms recognized four *Citrullus* spp

1. *Citrullus lanatus* (Earliar as *C. Vulgaris*)
2. *Citrullus colocynthis* (L) Schrad
3. *Citrullus ecirrhossus* Coryn
4. *Citrullus naudinianus* (Sond) Hook.,

1. ***Citrullus vulgaris/ Citrullus lanatus*:** It is annual and was originated in Southern Africa. Distributed in Egypt, South, west and central Asia, Leaves large broad, which are orbicular to triangular, ovate in shape, deeply 3-5 lobed or sometimes simple, medium in size. Flowers: Medium size, monoecious and short pedicels. Fruit: Medium to large size, thick rind and solid flesh with high water content, flesh colour red, yellow and white. Seeds: Ovate to oblong, strongly compressed, white or brown seed coats. Origin of cultivar: Cultivated watermelon originated in Japan, Bitter and non bitter wild race found in cape provinces (SA).
2. ***C. colosynthes**:* It is perennial, originated in North Africa, Rabot (Morocco). Compared to *lanatus,* leaves are small with narrow lobes hairy and grayish

in colour. It is monoecious and flowers are small, profuse blooming occurs in autumn when fresh vegetative growth occurs. Seeds are small and brown. Fruits are small, not exceeding 3 inch in diameter with rind and spongy flesh that is always bitter.

3. ***C. naudinianus***: It is perennial, vegetative characters differ from those of other species, where Leaves are deeply palmatifid and covered with dense fine hairs, tendrils simple, straight elongated or slightly curved at the apex. It is originated in South West Africa. Flowers are diocious, not formed until the second year of growth. Fruits are ellipsoid in shape, medium to large in size, thin rind and soft juicy flesh. Seeds are white and would not germinate under natural conditions.

4. ***C. ecirrhosus:*** It is perennial and originated in south West Africa. It closely resembles *C.colosynthis* in vegetative characteristics but leaves are more divided and covered with dense fine hairs, and have strogly recurved margins, tendrils absent, no flowers produced until the second year of growth. Fruit are sub globose with white flesh and bitter like *C.colosynthis.*

5. ***C. fistulosus****:* It is originated in India, annual, closely resembles *Cucumis* species and it will not hybridise with *Citrullus* spp and has different chromosome number (2n=24) of that of *Citrullus* spp (2n=22) and hence reclassified into *Cucumis* spp and then given separate genera status as *Praecitrullus*.

All four species of *Citrullus* could be crossed with each other successfully and F_1 seeds from almost all crosses germinated. F_1 seedlings grow normally and set fruit with good seeds. All these species have 2n=22 chromosome number (Shimotsuma, 1963, cytological studies)

General Botany: Water melon plant resembles *Cucumis* but stems are angular in cross section and leaves are cordate at the base and are pinnately devided in to 3 or 4 pairs of lobes. A non lobed (entire) leaved mutant was found and may be used as genetic marker (Mohr *et al.*, 1955). It has extensive root system but shallow, consisting tap root and many lateral roots growing within the top 2 feet of the soil. Early destruction of tap root (transplanting) may be advantages in getting superior yields (Emstrom, 1973). It is annual vine, but dwarfs forms exist and referred as bush types which are advantages. Bush types are as low as two feet and vine types are as more as 30 feet long. Dwarfing is primarily related to shortened internodes (Mohr, 1956).

Vine Types: There is delay in development of lateral branches with dominant single runner reaching several feet and branching initiated.

New dwarfing types: Multiple branching occurs simultaneously from crown of plant when it is quite young, which provide more potential bearing areas for concentrated fruit set (Mohr and Sandhu,1975).

Pollination: Naturally cross pollinated crop, despite large amount of self pollination and sibbing occurs naturally. This is demonstrated by use of marker gene (Mohr *et al.*, 1955).

Fruit: Fruit size greatly varied as demonstrated by fruit weight from 3 lb to 100lbs. Present day cultivars varies approximately weigh around 25lbs and future appears to be towards small size fruited cultivars. Fruit shape varies from long cylindrical to spherical intermediate types predominant. Rind colour vary from shades of green to white, some cultivars have striped rind and some have mottled rind. Greater part of fruit is edible flesh, which is mostly derived from the placenta where as *Cantaloupe* (musk melon) and squash are largely pericarp. The pericarp of watermelon is its hard outer rind.

Seeds: Size vary from very small (tomato seeds) to large pumpkin seeds. Seed colour white to black, red, green, brown and mottled colours also occur. Seed viability is 5 to 6 years.

Floral biology: Flowers are smaller and less showy than those of most of other species of cultivated *cucurbitaceae*. Flower located in leaf axils, most commonly singly. Most cultivar are monoecious, few older cultivar are andromonoecious (perfect flower and staminate flower). The pistillate or hermaphrodite flowers normally occur in every seventh (7th) leaf axil, the intervening axils being occupied by staminate flowers. The corolla is greenish- yellow, united in tube, deeply five lobed. Three stamens are inserted at the base of the corolla.

Pollination: Generally pollinated by honey bees (Porter, 1933). Andromonoecious cultivars do not confer the advantage of self pollination except with hermaphrodite flower. Hermophrodite flower must be visited by insects to effect the pollination. Andromonoecey has no advantage over monoecy in maintaining pure lines. Naturally favours cross pollination and consequently considerable genetic variations within a cultivar are maintained. Flower opens shortly after sunrise and remains open for a day. Pistillate and staminate flowers are in the axil, open on same day. Usually anther dehisce when corolla expands. Stigma is receptive throughout the day, but artificial selfing between 6-9 am gives maximum fruit set compared to later part of the day. High atmospheric humidity favours fruit setting. Big size ovary greater the chance of setting. Largest ovaries are usually found on flower near the tips of the most vigours branches of a plant. Unlike some other members of cucurbit family, the watermelon does not have flowering peaks or fruiting cycles. However, there is inhibitory influence produced by fruit already set, that reduces further fruit setting. For controlled pollination you should remove earlier set fruit.

Controlled Pollination: For pollination remove petals of male flower and brush sticky mass of pollen on anthers against the stigmatic surface of the pistillate flower. Prior to pollination unopened buds should be protected from insect visit. Use small screen cages or paper clip or scotch tape. After pollination protection must be replaced (Clip/tape). Can use plastic flags at each flower to be used. Different colours can be used to designate location of developing fruits, flower to provide pollen and pistillate buds not yet opened. Tags should be attached to the pedicel of pistillate flower to identify the pollen parent by indicating the date of pollination.

Breeding objectives

Global

1. Yield and quality

   Quality : 10% TSS, deep red flesh few small seeds.
2. Earliness (Pistillate/ hermaphrodite flower at early node).
3. Exploiting dwarfness.
4. Disease resistance (Anthracnose, PM,DM and virus)
5. Development of F_1 hybrids
6. Tough skinned fruits for long transport.
7. Development of seedless water melon

In India objectives not prioritized and need to be given importance with respect to following points.

- An early maturing variety : Pistillate/ hermaphrodite flower at early node.
- Tough skinned fruits for long transportation.
- Sweetness >10% TSS
- Fruits with fewer and small seeds and attractive deep red flesh.
- Resistance to diseases like PM, DM, Anthracnose and virus and insect pests.

Watermelon varieties are sbeing tested at different stations like Akola, Ludhiana, Durgapura, Hessarghatta, Vellanikkara, IARI New Delhi, Sabour, Navasari, Jabalpur and Ambajogai.

Genetic Resources: Durgapur and Faizabad centers made responses for collection and evaluation of germplasm of watermelon.

Cytogenetics: 2n= 22, C. *fistulosus* raised to new genus *Praecitrullus* due to its cross incompatibility with other species of *Citrullus* and differing chromosome number (2n=24).

Trait profile

1. **Yield and Quality:** Number of fruits and total weight is considered for selection, preference shifting towards small size fruits. This may favour mechanized harvesting. For small fruit size, need to look for thinner (but tougher) rind, firm flesh (Shipping durability) and small seed size (to avoid out of proportion with edible fruit). Colour of flesh where intense red preferred over pale red and pink, yellow flesh also preferred. Preference is flexible for fruit colour. Total soluble solids (TSS) reflecting sweetness should be high to catch Indian markets. Texture of flesh should be melting (or fine grained) flesh which is preferred over fibrous course grained form. Shape of fruit suited for packing and mechanized harvesting are preferred where consumers may not have preference.
2. **Earliness:** Earliest maturing cultivars of past have small fruits and it may not be disadvantage as now preference is towards small fruit size. Generally from setting to maturity require 30-35 days. This varies earliest 26 days to late type 45 days. Both environment and genotype control earliness.
3. **Exploiting dwarf habit:** Excessive vegetative growth clog up machinery and makes engineering equipment a difficult task (Mechanized harvesting). With the discovery of dwarfing genes in watermelon there was hope for cultivars suited for mechanized harvesting. It appears to be a feasible goal. Need to combine dwarfness with early maturity, high quality, increased production and other characters. It suit for home garden to save the space.
4. **Plant height:** No clear genetic information on vine size and vigour is available. dw2 dw2 : Multiple branching for crown, dw1 dw1: dwarf plant

dw1 dw1 (Dwarf plant) X DW1 DW1 (Multiple Branching plant)
DW2 DW2 ↓ dw2 dw2

F_1 normal vines
DW1dw1 DW2 dw2

F_2 Segregation ratio

9 : vines (Dw1- Dw2-)

3 : short internodes dwarf (dw_1dw_1 DW_2)

3 : multiple branches (Dw1-dw2dw2)

1: double recessive dwarf genes (dw1dw1dw2dw2) (Cv: Kengarden)

Totally new plant type (dw1dw1dw2dw2) appeared in F2 having very short internodes and crown branching.

5. **Leaf color** : Pale colour dominant to normal and 3 unlinked genes involved (Rhodes, 1986).
6. **Leaf shape:** lobed leaf shape is incompletely dominant to non lobed. Leaf shape with non lobed leaves governed by recessive gene 'nl'
7. **Fruit shape:** Elongated fruit incompletely dominant to spherical. Long fruit incompletely dominant over round fruit and is monogenic (Nath and Dutta,1973) and there are also report that round incompletely dominant over long (Brar and Nandapuri, 1974) and it depends on genetic source. Round fruited cultivar are susceptible to hallow heart. Elongated fruit tends to be gourd neck fruit and hence intermediate types avoid these problems. It has qualitative inheritance with one pair of allele (single gene) where F_2 segregated in 1 round: 2 intermediate: 1 elongated ratio and F_1 is intermediate and appears to be the case of co dominant intra-allelic interaction.
8. **Fruit skin colour:** Dark green dominant over light green involving single gene (Sachan and Nath, 1972)
9. **Fruit stripe colour:** Dark green stripes dominant to non striped with single gene control (Sachan and Nath, 1972). Striped green is recessive to solid dark green. Striped green is dominant to solid light green (DG>SG>LG). Greenish white mottling is determined by single recessive gene. Furrowing or grooving of rind recessive to smooth surface but no practical importance. Dark green melons seems to be more prone to sun burning resulting in to collapse of rind tissue. Single gene is involved and it is a good seedling marker.
10. **Rind colour:** Single gene determines intensity of green colour where dark green is dominant over light green.

Fruit rind colour, texture

Gene symbol	Description
F	Furrowed fruit surface, recessive to smooth
g synonymous D	light green (recessive to dark green)
g^s synonymous d^s	striped green skin (dominant to light green)
M	mottled skin
g^o synonymous c	Golden
e synonymous t	explosive rind

11. **Flesh colour:** Older cultivars large fruited and controlled by polygenes (25 genes involved) and now trend is towards small fruits (Poole and Grimball, 1945). Yellow flesh is recessive to (y/y) to red am dos controlled by single gene. Shade of yellow (Canary) dominant to pink flesh. Red flesh of *C. lunatus* (wf/wf) is recessive to white flesh colour (Wf/-) of *C. caolosynthis*). Red is dominant over yellow, canary yellow is dominant over pink and controlled by pair of genes (Sachan & Nath, 1972). Canary flesh (shade of yellow) dominant to pink & Intense red flesh colour inheritance is not studied.

    Fruit flesh colour segregation in F_2 is 9:7 ratio and hence it is having complementary gene action.

Genotype	Phenotype
Wf – Y –	White flesh
WfwfY–	Red flesh
Wfwf yy	Ref flesh
wfwf yy	Ref flesh
Yo	Orange flesh

Reports suggesting genetics of Fruit flesh colour

R syn Y	red flesh
y syn	(r) yellow flesh
wf syn (w)	white flesh dominant to red

12. **Flavour:** Not a clearly define trait, some off flavour termed as "Caramel" and some people will not recognize it and it is associated with intense red flesh and it is heritable and can be eliminated by selection. Flesh flavor is controlled by 'su' (synonymous Su^{Bi}) gene which is suppressor of bitterness. Bitterness is undesirable trait and associated with lines derived from *C. colosynthis* (gene transfer).
13. **Total soluble solids:** It also adds to flavor. Higher TSS greater preference, where more than 12% is preferred and less than 9% is not desirable.
14. **Seeds:** Quantity of seeds is heritable and consumers will not prefer large number of seeds and seed producers will not prefer too few seeds since, it leads to expensive seed cost. Best way is to reduce seed size.
15. **Seed size**: Small seed size is dominant over large (Sachan and Nath, 1972) where one gene is involved. In very small melons (3-5lb), a large number of large seeds result in a relatively low percentage of edible flesh.
16. **Seed colour:** There is strong prejudice against white seed colour which consumer associates with melon immaturity. Black seed seems to intensify

red flesh colour. Black is dominant over dark brown, red, dark orange yellow & dark orange. Dark brown is dominant over red, dark orange & dark orange yellow. Dark orange yellow is dominant over red (Sachan and Nath, 1972).

17. Seed coat colour: **w**: White coat, **t** : tan coat, **r**: red seed coat

18. Shape l: long seed s : Short seed

Gene Symbols table Adopted from Robinson *et al.*, 1976

Character	Number of genes	Types of gene action
Sex	Monogenic (AA)	Monoecious dominant to andromonoecious
Leaf lobing	Monogenic	Lobed dominant to non lobed
Dwarfness	Digenic (dw1, dw2)	Spreading dominant to dwarf bushy nature
Male sterility and glabrous leaf	Monogenic (msg, msg)	Male sterility and glabrus leaf are pleiotropic and governing by recessive gene
Fruit shape	-	Round partially dominant to long fruit
Fruit weight	Monogenic	Higher fruit weight incompletely dominant to lesser fruit weight
Seed number	Monogenic (NS)	High seed number dominant to low number
Rind thickness	Digenic (Rt, Rt2)	Thick incompletely dominant to thin
Days to first male flower	Monogenic (Dm)	Late dominant
Days to first female flower	Trigenic (Df, Df2, Df3)	Incompletely dominant
Growth rate	Trigenic (Gr1, Gr2, Gr3)	Incompletely dominant
Days to fruit ripening	Digenic (Dr1, Dr2)	Incompletely dominant
Rind toughness	Digenic (Tr, Tr2)	Toughness incompletely dominant to fragile (lesser toughness)

Gene symbol	**Description**
Ar-1	Anthracnose resistance to race 1
Fo-1	Dominant gene for resistance to race 1 of *Fusarium oxysporum* sp. *niveum*
pm	Powdery mildew susceptibility: recessive gene
Fwr	Resistance to fruit fly: dominant gene

19. Red pumpkin beetle resistance: Resistance dominant, Monogenic (Vashishtha and Choudary, 1972)

20. Sex expression : Monoecious (A/-) dominant to Androeciuos (a/a) where 6 staminate flower and 1 pistillate flower occur, there are lines where ratio is 3 (staminate) : 1 (pistillate) in same plant. Male sterility also reported (gms/gms), mutant was glabrous (Watts, 1962)

Quantitative traits: Genetics of traits varied from population to population and hence, varied kind of genetics reported for the same trait.

TSS : Additive (Brar and Nandpuri, 1974), Add & dominant (Sidhu *et al.,* 1977)

Fruit shape index: Partially dominant and has non allelic interaction (Dhaliwal, 1982)

Earliness: Dominant, dom X dom epistatis, duplicate and complementary types of epistatis reported (Sharma and Choudary, 1988)

Fruit weight: Dominant and dominant X dominant epistatis reported (Sharma nad Choudary, 1988).

Storability (Durability) : Select for thick, tough and flexible rinds (multiple factors involved) and firm flesh (heritable but mode of inheritance not studied). With evolution of small melons suited for packing reduces scope for durability.

Breeding Methods Ploidy Breeding

Seedless triploid hybrids: Diploid pollen on triploid stigma stimulates parthenocarpy, but ovules fail to develop. Diploids treated with colchicines get Tetraploid and this when crossed with Diploid pollinator results in Triploid (Reciprocal crosses not successful). Tetraploid produces small number of seeds than diploids and hence expensive to maintain tetraploid. Therefore triploid seed cost is 20 times of that of open pollinated seeds. Additional disadvantages in triploid are seeds difficult to germinate, removal of seed coat is recommended by Japanese and require high temperature (86° F). For Indian condition it may not be a problem. Therefore triploid watermelons were not economically feasible in USA in earlier days. The first report of triploid (seedless) watermelon in the United States' was described in the literature in 1951 (Kihara, 1951). However, watermelon cultivars for consumption prior to 1990 were primarily comprised of diploid (seeded) cultivars that produced fruits which were generally at least 10 kg. Seeded "ice box" melon such as Mickylee and Minilee were introduced in 1986, and gained increased popularity. Only in early l990s did seed for seedless watermelon cultivars become more readily available commercially. Improved seed germination and production practices, as well as increased market demand, have resulted in increased seedless watermelon production. During 2007, in United States, seedless water melon production comprises nearly 70% of watermelon shipments. Triploid watermelons are being sold since from 1980s and in 2004 comprised nearly two thirds of the watermelons in US. In 2003, mini seedless watermelons averaging about 2.5 Kg became available to the American consumer. These initially introduced mini seedless watermelons are sold under the brand names of Pure Heart and Bambino and were developed by Syngenta seeds and Seminis Inc., respectively. These brand name mini seedless watermelons only be grown by contract growers. Other seed companies have developed their own mini seedless watermelon cultivars.

Table: Triploid mini watermelon seed sources and descriptions, 2004.

Entry No.	Cultigen	Company	Description
1	Betsy (8103)	Nunhems USA	Distinct, narrow, dark green stripes on light green background; round to slightly oval; uniform shape and size
2	Bobbie (8103)	Nunhems USA	Distinct, narrow, dark green stripes on light green background; mostly round to slightly oval; overall uniform shape and size
3	HA 5109	Hazera Seeds	Solid, dark green; primarily round fruit; golden yellow ground spot appears when ripe; fairly uniform in shape and size
4	HA 5117	Hazera Seeds	Some distinct; some indistinct medium width, dark green stripes on medium green background; slightly oval shape; overall uniform shape and size
5	HA 5130	Hazera Seeds	Distinct, narrow, dark green stripes over light green background; oval fruit shape; shape is uniform; size is variable
6	HA 5133 (Mielhart)	Hazera Seeds	Distinct, narrow, dark green stripes on light green background; oval shape fruit; uniform shape and size
7	HA 5135	Hazera Seeds	Distinct, narrow, dark green stripes on light green background; oval fruit shape; uniform fruit shape; variable fruit size
8	HA 5138	Hazera Seeds	Distinct, narrow, dark green stripes on light green background; slightly round with fairly uniform shape and size
9	HA 5144	Hazera Seeds	Distinct, narrow, dark green stripes on light green background; slightly more oval than round fruit shape; large fruit; overall, fruit too large
10	HA 6007 (Xite)	Hazera Seeds	Distinct, narrow to medium-wide dark green stripes on a medium-dark green background; slightly oval fruit shape; overall uniform fruit shape and size
11	HA 6008 (Extazy)	Hazera Seeds	Indistinct,medium to wide dark green stripes on medium green background; slightly oval to round; assymetrical fruit on 1st harvest; more uniform on 2nd harvest
12	HA 6009	Hazera Seeds	Indistinct,medium to wide dark green stripes on medium green background; oval to round fruit shape; fairly uniform shape and size
13	Mini Yellow	D Palmer Seed	Solid, dark green; round fruit shape; bright yellow ground spot when ripe; large fruit size for mini melons; some fruit too large; uniform shape

14	Mohican	Southwestern Seed Co.	Indistinct, medium wide, dark green stripes or medium green background; primarily oval fruit shape; some asymmetrical fruit
15	Petite Prefection	Syngenta Seed Ine.	Distinct, very narrow, dark green stripes on light green background; slightly oval fruit; uniform shape and size; excellent mini melon size
16	Petite Trear	Zeraim Gedera Ltd.	Distinct, narrow, dark green stripes on light to medium green background; mainly oval fruit;with some round fruit shape; size is very variable
17	Precious Petite	Syngenta Seed Ine.	Distinct, narrow, dark green stripes on light green background; round to oval fruit shape; uniform size and shape; many fruit have mini melon is very variable
18	PS 4911714	Seminis Ine/ 6-L Farm	Indistinct, solid; medium to dark green; slight with golden yellow ground spot; uniform; shape variable size
19	RWT 8149	Syngenta Seed Ine.	Distinct, very narrow, very dark stripes on background; primarily round fruit shape; some variable in fruit size
20	RWT 8154	Syngenta Seed Ine.	Indistinct, extermely nerrow, broken medium stripes on light green background; round to oval fruit shape; golden yellow ground spot ripe uniform fruit shape and size; good mini size
21	RWT 8155 (Bibo)	Syngenta Seed Ine.	Distinct, very narrow, dark green stripes on light green background; oval fruit shape; uniform shape and size; good size for mini melon
22	RWT 8162	Syngenta Seed Ine.	Distinct, extremely narrow, light green stripes disappear when fruit ripens; rind turns pale yellow ripers; round to oval fruit shape; overall shape size are fairly uniform
23	Solitaire	Seedway Ine.	Indistinct, medium dark green stripes on light medium green background; round to slightly fruit shape; uniform shape; variable fruit size
24	SWX 9001	Southwestern Seed Co.	Distinct, very narrow, dark green stripes on light green background; oval uniform fruit shape; size is uniform; but generally too large for a melon; interior quality is not commercially acceptable
25	Valdoria	Nunhems USA	Solid, dark green rind; primarily round fruit bright yellow ground spot when ripe; variable size; uniform fruit shape
26	Vanessa	Nunhems USA	Solid, dark green rind; round fruit bright yellow ground spot when ripe; fruit size and are fairly uniform; many fruit are too large

At IARI crossed Tetra 2(from USA) with Pusa Russel (diploid) and developed seedless triploid hybrid Pusa Bedana (Sheshadri *et al.*, 1972).

Steps

1. ***Development of tetraploids:*** Application of 0.2 to 0.4% colchicines during morning and evening two times to the growing point of young seedling for 4-6 days and prior to this soaking of seeds in 0.4 % solution for 24 hour. This can result in to doubling of chromosomes.
2. ***Detecting tetraploids :*** Tetraploids cans be identified based on certain markers like large pollen size and large seed size and different rind pattern. Identified tetraploids should be stabilized and further used for development of triploids.
3. ***Development of triploids :*** Crossing of tetraploids (4x) with help of pollen from diploid (2x) parent.
4. ***Identification of good specific combiners:*** Need to develop more number of crosses using different parents and crosses need to be assessed for identification of good specific combiners and then the triploid hybrid can be evaluated for various quality parameters before releasing it for commercial cultivation.

Heterosis: Heterosis is reported for yield (Nath and Dutta, 1970), earliness and TSS (Nandpuri et al, 1974), Yield and number of fruits (Gill and Kumar, 1988). Each fruit produces large number of seeds (approximately 225) and F1 hybrids are feasible but require sufficient heterosis for yield.

Development of F1 hybrid cultivers: Data to establish yield superiority of F1 hybrids are lacking and need to explore inbreds extensively for development of commercial hybrid. One of the most sensational F1 hybrids among vegetables is seedless triploid watermelon. Unfortunately it has led to high cost of F1 seed and its poor germination. Improvement to solve this problem is very much essential. There is no inbreeding depression.

Advantages of F1 Hybrids

Uniformity

Dominant gene controlled Disease resistance can be exploited

Intermediate fruit size

Less gourd necking

Less hallow heart

Resistance Breeding: Fusarium wilt, anthracnose, Gummy stem blight, Powdery Mildew, Downey Mildew and Viruses are important diseases in watermelon.

***Fusarium* wilt** resistance in Watermelon is governed by multiple factors mostly recessive. Therefore development of resistant variety is slow process and races of *Fusarium* also occur. *Fusarium* wilt is caused by *Fusarium oxysporum* f. sp *niveu.*

Anthracnose: Diseases is caused by *Glomerella singulata* var *orbiculare*. High humidity favours disease development. Air borne spores attacks both plants and fruits. Single dominant gene governs resistance, several races exists.

Powdery Mildew: The 'pm' gene is susceptibility to Powdery Mildew.

Breeding Programmes

Florida Breeding Programme for Disease resistance and quality (Crall and Elmstrom, 1979): The gene W5 offers superior quality fruits and is reported in the cultivar Texas W5. The cultivar Summit possess high type wilt resistance. The cultivar 'Wilt Resistant Peacock 132 (WRP) possesses multiple recessive factors, intense red colour, firm flesh and tough rind. The cultivar 'Fairfax' possesses anthracnose resistance (Single dominant gene). Using all these cultivars breeding procedure was followed as mentioned in the flow chart.

Florida Breeding Program [Crall and Elmstrom (1979)]

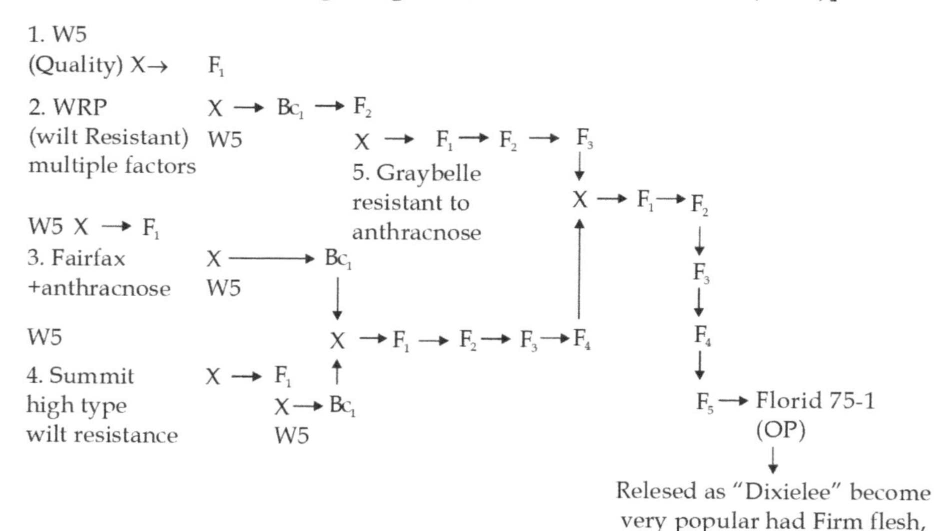

Florid-75-1 released in the name of Dixielee and become very popular. It was superior with respect to flesh firmness, thin rind, TSS (11.2%), flesh colour dark red and had high consumer preference. It required totally 14 years to achieve objectives.

Kentuky Watermelon Breeding Programme (Dwarf types and Bush types): To develop commercially acceptable dwarf types breeding program is suggested. Mutant from Cv. Desert King carried gene 'dw1dw1' and has yellow flesh, yellow rind and very late maturity and disease susceptibility. Vine type parents selected to overcome these disadvantages and advance line designated as W-45 was selected. In 1962, dwarf mutant was reported in Japan Cv. Asahi Yamato by Shimotsuma Mohar H.C (breeder). Along with W-45 line he moved from Texas to Kentuky and carried out the work.

Japanese dwarf mutant (dw2dw2) X W-45 (dw1dw1)

$F_1 \rightarrow F_2$

In F2 new double recessive dwarf (dw1dw1 dw2dw2) appeared and it was superior to single gene dwarf.

Double recessive dwarf was crossed with New Hampshire Midget and material carried to F5 generation.
F1 → F2 → F3 → F4 → 25 F5 lines selfed,

Satisfactory uniformity for quality, expected seed colour and size obtained in F5, subsequent selfing resulted with desirable uniformity for all characters and named the line as Kengarden and released for cultivation.

Salient Breeding Achievements

Varieties

Caltham Gray, Summit, Shipper and White Hope are resistant to *Fusarium oxysporum* f. sp. *niveum*

Fairfax: Resistan to Alternaria cucumerina

Alena is a tetraploid from diploid Sugar Baby.

Durga Dura: Local cultivar

AHW19: Suited for hot arid region, 3 to 3.5 fruits /vine, dark green strips, 4 kg/ fruit, dark pink flesh, 8 to 8.4 % TSS, 460 to 500 q/ha.

AHW65: Suited for hot arid region, 3 to 4 fruits /vine, green strips, 2.5 to 3 kg/ fruit. Pink and solid flesh, TSS is 8 to 8.5% and yields 375 to 400 q/ha

Asahi Yamato: It is a midseason Japanese introduction producing medium sized fruits averaging 6 to 8 kg. The rind colour is light green with deep pink flesh, TSS 11-13%. Fruits ripen in 95 days. Fruit is round and has non-striped skin.

Sugar Baby: An early season American introduction where fruit is slightly smaller in size weighing 3 to 5 kg, round, having bluish black rind and deep pink flesh, TSS 11 to 13% and possess small seeds and yields 150 q/ha in 85 days.

Pusa Bedana: (IARI): It is a seedless obtained as a hybrid between Tetra-2 of the USA and Pusa Rassal, a local purified type. Flesh is deep pink and sweet.

New Hampshire Midget (IARI): It is an early variety introduced from the USA and fruit is small (1.5-2 kg), oval in shape, skin colour bright green with dark green lacerations and red flesh inside. It takes 28 to 30 days from the date of pollination to ripening. It is an early home-garden variety.

Arka Jyothi (IIHR): A mid season F_1 hybrid (lIHR-20 from Rajastan x Crimson sweet from USA) with round fruits weighing 6 to 8 kg. The rind colour is light green with dark green stripes and flesh colour crimson and 11 to 13% TSS and yields 840 q/ha and has low seed content.

Arka Manik (IIHR): Evolved by crossing IIHR 21 with Crimson sweet. Fruits are round to oval with green rind and dull green stripes. The flesh is deep red, very sweet (12.15% TSS). The average fruit weight is 6 kg and has low seed content. It stands well in transport and storage. It yields 600 *q/ha. It is* resistant to Powdery Mildew and Downy Mildew and anthracnose diseases.

Improved Shipper: An introduction from USA having big fruits weighing 8 to 9 kg and fruit is dark green with moderate sweetness (8-9% TSS) with characteristic tripartite blossom end. It yields 300 q/ha.

Durgapur Meetha: A late cultivar maturing in 125 days, fruits round with light green. Rind is thick with good keeping quality, flesh sweet, TSS around 11 % with dark red colour, average fruit weight is 6 to 8 kg. Seed is with black tip and margin. Released by ARS, Durgapur, Rajasthan.

Durgapur Kesar: A late cultivar, fruit weight is 4 to 5 kg. Skin is green with stripes, flesh yellow in colour, moderately sweet, seeds are large. Released by ARS, Durgapur, Rajasthan.

There are several cultivars, locally grown which are named after the region in which they are grown, such as Farrukhabadi, Moradabadi, Faizabadi of UP.

Foreign cultivar: Charleston Grey-USA. All jubilant and all producer of -USA. Red-N-sweet-USA. Wuchazao-China. Moodeungansoobak-Korea. Hungaria (H-8) and Napsugar (Sun beam) - Hungary and Vodolei-Russian.

Sugar Baby: This is a popular American variety, introduced at IARI, New Delhi. Plants are with medium long veins. Fruits are round with black rind, deep pink flesh with 11-13% T.S.S., brown small seeds, weight varies from 2-5 kg; gives an average yield of 15 t/ha in 85 days of crop duration. Recommended for release and cultivation in 1975 for the zones V (Chhattisgarh, Orissa and A.P.), VII (M.P. and Maharashtra) and VIII (Karnataka, Tamil Nadu and Kerala).

Thar Manak: Fruits 2.65-4.2 kg, ready for harvesting 70-80 DAS. No. of fruits 2.59 to 4.22 per plant, yields 50-80 t/ha, suitable for summer and rainy seasons. Fruits are free from cracking under extremes of high temperature in summers. Fruits are oblong-round having dark green-green stripes on the smooth rind. Flesh is red, solid (firm) and granular and has good taste and sweetness (9.5-11.2 % TSS). Seeds are big, bold and blackish in colour. It has low in seed content (160 to 266/fruit).

Arka Muthu: High yielding dwarf vine of 1.2 mt, vine length, shorter internodal length and early maturing type (75-80 days). It has Round to oval fruits with dark green stripes and deep red flesh. Average fruit weight is 2.5-3 kg with T.S.S ranging from 12 to 14 %. Fruit yield 55 to 60 t/ha.

Arka Aiswarya: High yielding F_1 hybrid. Green with Dark green deeply lobbed foliage, round to oval fruit, Dark green with light green broken stripes, red flesh, with TSS of 13-14% (brix), average fruit weight 7.5kg with 1-2 fruit per vine. Duration is 95-100 days. Fruit yield 75 to 80 t/ha, red flesh, crispy, delicious, juicy and very good taste. Good keeping and transport qualities.

Arka Akash: High yielding F_1 hybrid. It has oblong fruit with red flesh, TSS of 12-13% (brix), average fruit weight 6.5kg with 1 fruit per vine. Duration is 90-95 days. Fruit yield 65 to 70 t/ha and has red flesh, juicy and very good taste and has got good keeping and transport qualities.

Arka Madhura: Triploid seedless watermelon variety and high yielding with 50-60 t/ha, T.S.S 13-14 %. Unique type, sweet, juicy and fully seedless , longer shelf life and transport quality, suitable for year round production under protected condition.

9

Ridge Gourd and Sponge Gourd

Ridge gourd or Ribbed gourd (*Luffa acutangula* Rox b. L) and Sponge gourd or Smooth gourd (*Luffa cylindrica*) have diploid chromosome number 2n=2x=26. Luffa has old world origin in Subtropical Asian region including India (Kalloo, 1993). These two species are inter-crossable. F_1 plants generally, intermediate between the parents and showed various irregularities like, univalants, rings and chains of 4 chromosomes, chromatids bridges and fragments at metaphase and thus the species are not easily crossable and the F_1 plants appear to be of not much practical value. The genus is monoecious annual vine, tendrils are branched, flowers are yellow, anthers are free and pistil has three placentas with many ovules and stigma are three and bilobate. Fruits oblong or cylindrical and rind becomes dry on maturity.

Objectives of Breeding

1. Earliness
2. High female to male sex ratio
3. Uniform thick cylindrical fruits free from bitterness
4. Tender, nonfibre fruits for longer time
5. High fruit yield and more fruit number with more fruit weight
6. Resistant to powdery mildew and insects

Genetics: Singh *et al.*, (1948) reported that two loci A and G determine sex expression in *Luffa acutangula.*

Monoecious X Hermoprodite

AAGG ↓ aagg

F_1 (Aa Gg) → F_2

9: Monoecious A_G_ 3: Andromonoecious A_gg

3: Gynoceious aaG_ 1: Hermaphrodite aagg

Bitterness governed by single dominant gene 'Bi' (Thakur and Choudhary 1966).

Monogenically inherited traits

1. Corolla colour: Orange yellow of *Luffa cylindrical* dominant to Lemon yellow of *Luffa acutangula*
2. Fruit surface: Ridged of *Luffa acutangula* is dominant.
3. Seed surface: Pitted dominant to Non pitted in *Luffa acutangula*

Breeding methods suggested and followed

1. Inbreeding and selection
2. Pedigree method
3. Bulk method
4. Back cross method
5. Single seed descent method
6. Heterosis breeding

Varieties Developed

Ridge gourd: There are many released varieties in ridge gourd viz., Co-1, Pusa Nasdhar, Satputia (hermaphrodite), Pant Torai 1, Konkan Harita, Punjab Sadabahar, etc.

1. **Pusa Nasdar (Kalitori):** It is developed at IARI, New Delhi. It is selection from local material of MP and suitable for summer and rainy seasons. It is moderatly early and posses club shape, light green long fruits. It produce 15 to 20 fruits per plant and yields 15 to 16 t/ha.
2. **CO-1:** It is developed at TNAU, Coimbatore.

   Growth habit Moderately vigorous

   Days to first fruit harvest : 55 days after sowing and it is completed in 125 days

   Fruits per plant : 10-12

   Fruit size : 60-75 cm long

   Fruit girth : 30 cm

   Fruits per kg : 3-4
3. **Pant Torai-1:** It is developed at GBPUAT, Vegetable Research Centre, Pantnagar. It is pure line selection from inbreds of indigenous germplasm

   Year of release : 1999 by UPSVRC

Area of adaptation	: Suitable for Northern plains
Days to first picking	: 65 days after sowing
Fruit shape	: Club shaped
Fruit size	: 15-20 cm long
Vine length	: About 5 m long
Yield	: 100 q/ha

4. **Hissar Kalitori:** It is developed at HAU, Hisar, Haryana.

Year of release	: 1995 at state level
Maturity	: Early
Fruit size	: Long, thin and straight
Yield	: 80-100 q/ha
Miscellaneous	: Tolerant to powdery mildew and suitable for rainfed areas

5. **Haritham:** It is developed at KAU, Kerala. It is developed through selection.

Year of release	: 2000 by State Seed Sub-Committee
Area of adaptation	: North zone of Kerala
Fruit length	: 46-47 cm
Fruit weight	: 650 g
Crop duration	: 95 days
Yield	: 132 q/ha
Miscellaneous	: Large light green, cylindrical fruit with typical ridges and tapering towards the base.

6. **Satputia:** It is a cultivar of Bihar, which bears hermaphrodite flowers.

Area of adaptation	: Mainly grown in Uttar Pradesh and Bihar
Fruit shape	: Small
Fruit colour	: Pale green
Fruits per cluster	: 5-7
Yield	: 200-250q/ha

7. **Konkan Harita:** It is developed at Kokan Krishi Vidyapeeth, Dapoli (Maharashtra). It is developed through selection.

Days to first harvest	: 45 days after sowing
Fruit colour	: Dark green
Fruit size	: Long (30-45 cm) and tapering at both the ends
Fruits per vine	: 10-12

8. **IIHR-8:** It is developed at IIHR, Bangalore.

Days to first fruit harvest	: 55 days after sowing
Fruit shape	: Round
Fruit colour	: Green
Fruit weight	: 100 g
Fruits per vine	: 40-45
Resistance	: Moderately resistant to downy mildew
Yield	: 400 q/ha
Miscellaneous	: Fruits having good keeping quality specially for stuffing purposes.

9. **Punjab Sadahabar:** It is developed at PAU, Ludhiana.

Sowing time	: This variety can be sown from May-July
Plant characteristics	: Plants are medium sized with dark green leaves
Fruit characteristics	: Fruits are long, 3-5 cm thick, slim, green, ridged, tender, slightly curved and rich in protein
Sowing time	: Suitable *for kharif* and summer seasons
Fruit colour	: Dark green
Fruit weight	: 300 g
Yield	: 280-300q/ha

11. **Kalyanpur Dharidar:** It is developed at Vegetable Research Centre, Kalyanpur, Kanpur.

Maturity	: Early
Fruit colour	: Light green
Yield	: 400 q/ha

12. **Arka Sumeet:** It is developed at IIHR, Bangalore. It is developed from a cross of Ill-JR-54 X IIHR- 24 followed by pedigree selection.

Year of release	: Released in 2001 by AICRP (VC) workshop
Area of adaptation	: Karnataka
Maturity	: Early
Days to first harvest	: 52 days after sowing
Fruit colour	: Light green
Fruit length	: 50-65 cm with prominent ridges
Fruit shape	: Cylindrical
Fruit weight	: 380 g

Fruits per vine	: 13-15
Vine length	: 4.5 m
Fruit yield	: 500 q/ha
Miscellaneous	: Good transport and cooking quality

13. **Arka Sujatha:** It is developed at IIHR, Bangalore. It is developed from a cross of IIHR-54 X IIHR-18 followed by selection.

Year of release	: 1996 by SVRC
Area of adaptation	: Karnataka
Fruit size	: Medium
Fruit colour	: Light green
Fruit length	: Medium long (35-45 cm) with prominent ridges
Fruit shape	: Cylindrical
Fruit weight	: 350 g
Vine length	: 5.5 m
Fruit yield	: 450-500 q/ha
Miscellaneous	: Good transport and cooking quality

14. **Swarna Manjari:** It is developed at HARP, Ranchi.

Area of adaptation	: Recommended for cultivation in zone IV
Days to first harvest	: 65-70 DAS
Fruit characteristics	: Fruits are elongated, medium sized, highly ridged, green and soft pulps contain less fibre
Resistance	: Tolerant to powdery mildew
Yield	: 180-200 q/ha in 140-150 days

15. **Swarna Uphar:** It is developed at HARP, Ranchi. It is pedigree selection.

Area of adaptation	: Recommended for cultivation in zone IV
Days to first harvest	: 65-70 DAS
Fruit characteristics	: Fruits are elongate, medium sized (200g), weak ridge with soft pulp and less fibre
Yield	: 200-300q/ha in 140-150 days

15. **CO-2:** It is developed at TNAU, Coimbatore. It is selection from local germplasm.

Year of release	: 1984
Fruit characteristics	: Green, very long, fleshy with avg. wt. of 0.8-1.0 kg

No. of fruits/vine	: 8-10
Yield	: 250q/ha in 120 days

16. **PKM-1:** It is developed at TNAU, Coimbatore. It is induced mutant from the type H-160.

Year of release	: 1980
Fruit characteristics	: Dark green, 60-70 cm long, with shallow grooves
Yield	: 250-280q/ha in 160 days
Other varieties	: Konkan Harit, Haritham, Local Yard Long.
Sponge gourd	: There are few released varieties in sponge gourd viz., Pusa Chikni, Kalyanpur Chikni (Chadha and Lal, 1993) and Pusa Supriya.

Sponge gourd

1. **Pusa Chikani (Ghia Torai):** It is developed at IARI, New Delhi. It is selection from local material collected from Bihar.

Sowing time	: Spring-summer and rainy season
Days to 50% flowering	: 35 days after sowing
Fruit colour	: Dark green
Fruit shape	: More or less cylindrical
Fruits per plant	: Prolific, 15-20 fruits per plant
Yield	: 150-165 q/ha

2. **Pusa Supriya:** It is developed at IARI, New Delhi. It is suitable for spring-summer and kharif seasons.

Days to first picking	: 52-53 days in summer and 45 days in *kharif* season
Fruit colour	: Pale green
Fruit shape	: Straight and slightly curved at the stem end, narrow green linings and spots, skin very thin
Fruit size	: Medium long (15-20 cm)
Fruit weight	: 110 g
Fruits per vine	: 12-16
Yield	: 80-90 q/ha during spring - summer season and 100410 q/ha during kharif season

3. **Kalyanpur Hari Chikni:** It is developed at Vegetable Research Centre, Kalyanpur, Kanpur.

Year of release	: 1982
Area of adaptation	: Uttar Pradesh
Growth habit	: Creeper
Fruit characteristics	: Smooth, dark green, medium sized thIn fruits
Duration of crop	: 125 days
Resistance	: Fruit fly and soft rot
Yield	: 100-125 q/ha

4. **Azad Taroi-1:** It is developed at CSAUAT, Kanpur.

Year of release	: 2001
Area of adaptation	: Uttar Pradesh
Maturity	: Early
Growth habit	: Creeper
Fruit characteristics	: Green and smooth
Yield	: 125-150q/ha

5. **Pusa Sneha:** It is developed at JAR!, New Delhi. It is suitable for summer and kharif seasons.

Fruit colour	: Dark green
Fruit length	: 20-25 cm long
Days to first harvest	: 40-50 days after sowing
Yield	: 120 q/ha
Miscellaneous	: Suitable for long distance transportation and can withstand high temperature during the summer season

6. **Phule Prajakta:** It is developed at MPKV, Rahuri, Maharashtra. It is selection from local germplasm.

Sowing time	: Suitable for summer and *kharif* seasons
Fruit colour	: Green
Fruit size	: Medium
Yield	: 150 q/ha

7. **Rajendra Nenua-1:** It is developed at BAC, RAU, Sabour, Bihar.

Fruit colour	: Greenish white
Fruit size	: Long

Fruit characters	: Smooth and thick
Resistance	: resistant to fruit fly and fruit rot
Yield	: 250 q/ha

8. **Kashi Divya:** It is selection from a local landrace and has high TSS and yields 250 q/ha and released for UP, Jharkhand, Bihar and Punjab.

10

Bitter Gourd

It is rich in iron and Vit –C and native of tropical Asia particularly eastern India and Southern China (Sheshadri 1986 Mini Raja *et al.,* 1993).

Botany: Monoecious annual, staminate flowers small, yellow and borne on slender pedicel. Pistillate flowers solitary have small pedicel, flowers are yellow, leaves segmented pentamerous. Filaments 3, two are bilocular and one is unilocular. There are three short styles ended with three bilobed or divided stigmas.

Chromosome 2n=22: 6-centromere in middle, 3-submedian, 2-subterminal (Vargheuse, 1973)

Momordica dioica: 2n=28

Momordica cochincniensis: Sweet gourd of Assam

Germplasm

KAU-Vellanikara, IIHR-Bangalore, NDUAT-Faizabad, CSAUAT-VRS, Kalyanpur, BPUA&T-Pantnagar and IARI- New Delhi are the major sources of bitter gourd germplasm.

Objectives of Breeding

1. Early fruiting
2. High female to male sex ratio
3. Whitish green to glossy green fruit colour depending upon consumer preference
4. Less ridged fruit surface
5. Thick fruit for stuffing
6. Slow seed maturation in the fruit.
7. High yield.
8. Resistant to red pumpkin beetle and fruit fly

Genetics

Nath *et al.*, (1973): Monogenic traits

Fruit colour: green dominant to white

Seed colour: dark brown dominant to whitish brown

Seed size: small size is dominant to large

Breeding Methods

1. Inbreeding of land races and individual plant selection
2. Pedigree method
3. Bulk method
4. Backcross method
5. Heterosis

Varieties

1. **Pusa Do Mausami**: Selection from local collection and suites for both summer and rainy season. Fruits dark green, long (18cm), medium thick, club shaped 7-8 continues ridges, 8-10 fruits weigh 1kg.
2. **Arka Harit:** Selection from local germplasm from Rajasthan (1973). Fruits spindle shaped, glossy green, smooth rind and thick flesh. 120q/ha in 120 days.
3. **Pusa Vishesh:** dwarf vine, glossy green fruits, medium long, thick. Released from IARI New Delhi, 55-60 days maturity
4. **Kalyanpur Baramasi:** CSAUAT-VRS, Kalyanpur, Kanpur vigorous creeper, long fruits, thin tapery.100-125q/ha in 120. Tolerant to fruit fly and mosaic (for UP).
5. **Kalyanpur Sona**: CSAUAT-VRS, Kalyanpur, Kanpur fruits medium long, thickness, 110-125q/ha in 120 days. Tolerant to fruit fly and mosaic (for UP).
6. **Pant Karela 1**: Selection from inbreds of indigenous germplasm at pantnagar released in 1999 by UP . Fruits are thick, 15 cm long. Yield is about 150q/ha.
7. **Coimbatore Green**: It is a local type selected at TNAU, Coimbatore. Fruits are extra long, upto 60 cm and dark green weighing 300-400 g. yield potential is 180 q/ha.
8. **Phule Green**: Developed by pedigree method from a cross of Green Long x Delhi local at MPKV, Rahuri. Fruits are dark green, 25-30 cm long. Yield :200 q/ha in 150-180 days.

9. **Priya (MC 23):** KAU released variety and is high Yielding (20 -30 t/ha), has long green and spiny fruits with white stylar end and average fruit weight is 235 g and average fruit length is 39 cm.
10. **Preethi (MC 84):** KAU released variety and is high Yielding (15 t/ha) with attractive, white, medium sized, spiny fruits and average fruit weight is 310 g and average fruit length is 30 cm.
11. **Priyanka (MC 23):** KAU released variety and is high Yielding (28 t/ha) with large, spindle shaped, white fruits with smooth spines and average fruit weight is 300 g and average fruit length is 25 cm.
12. **Kashi Urvasi:** This variety has been derived from the cross IC-85650B x IC-44435A, having dark green and long fruits, mild projection, length 16-18 cm, fruit weight 90-110 g and yield 200-220 q/ha. This is suitable for cultivation under both rainy and summer seasons. This has been identified by UP State Horticultural Seed Sub Committee and notified during the XII meeting of Central Sub Committee on Crop Standard Notification and Release of Varieties for Horticultural Crops for the cultivation in U.P., Punjab, and Jharkhand.

11

Bottle Gourd

Bottle gourd (*Lageneria siceraria*) 2n=2x=22 is also known as white flowered gourd. Most important vegetable of ancient China.

Origin : According to De Candolle (1882) Bottle gourd has been found in wild form in South Africa and India. However, Cutler and Whitaker (1962) are of the view that probably it is indigenous to Tropical Africa on the basis of variability in seeds and fruits. This species appear to have been domesticated independently in Asia, Africa and the New world (Heiser, 1973).

Botany: It is monoecious, annual vine with soft pubescence. Leaves are cordate ovate to reniform-ovate, Flowers are white, solitary, open at night and has hairy ovary. Fruit shape is cylindrical, oval egg shaped, club shaped, round and other shapes also appeared.

Genetics of Important Traits

Bitterness: Single dominant gene (Path and Singh, 1950)

Fruit colour: Monogenically controlled (Kalloo, 1993)

Two major genes are involved for fruit shape (Path and Singh, 1950). Single gene for fruit colour involved (patchy v/s white) and reported by Path and Singh (1950). Monogenic recessive inheritance for andromonoecious sex form is reported by Singh *et al.* (1996). Hermaphrodite sex form is characterized by long corolla (3.5-4.5cm instead of 2.5-3.5cm in normal). In hermaphrodite condition fruits bears prominent blossom scar. Long fruit conditioned by AA genotype and Round fruit is due to aa genotype where Partial dominance is reported (Kushwaha, 1996).

Objectives of bottle gourd improvement

1. High yield
2. Greater fruit number
3. Greater fruit weight
4. High Female: Male flower ratio
5. Earliness (appearance of pistillate flowers at early node)

6. Round, long, club shaped fruits with sparse hairs on skin
7. Non-fibrous flesh at edible stage
8. Non- bitter fruits

Breeding Methods

1. Inbreeding of land races and individual plant selection
2. Pedigree method
3. Bulk method
4. Backcross method
5. Heterosis breeding: - Pusa Meghdhoot (Long fruited) is a F1 hybrid where 75% heterosis reported for yield. Pusa Manjari (Round fruited) is a F1 hybrid where 106% heterosis for fruit yield is reported.

It is a highly cross pollinated crop where isolation distance for seed production can be minimum 1000m.

Insect Resistance

Red pumpkin beetle is an important pest where cotyledon leaf damage is observed. Aphids are serious problem at seedling stage. Fruit fly is very devastating insect where significant percentage of fruit damaged is reported. There is need for crop improvement to impart resistance to these insects with improved yield and quality traits.

Disease resistance

1. PI-271353: Resistant to CMV, squash mosaic virus, tobacco ring spot virus, tomato ring spot virus and WMV.
2. Doodhi Long Green and Three Feet Long Green: Moderately resistant to five isolations of *Sphaerotheca fulisinea.*
3. Renshi: (Thaiwan variety): It is highly resistant to Fusarium wilt. Seed protein electrophoresis used for identification of varieties (Upadhya, 1995).

Breeding Achievements : Varieties

Hybrid-Kashi Bahar : This is a long fruited hybrid with green vine and vigorous growth, fruit straight, light green, length 30 to 32 cm and average weight is 780 to 850 g and yield is 500-550 q/ha. It is suitable for rainy and summer season cultivation. It is tolerant to anthracnose, downy mildew and *Cercospora* leaf spot under field conditions. This has been identified by UP State Horticultural Seed Sub Committee & notified during the XII meeting of Central Sub Committee on Crop Standard Notification and Release of Varieties for Horticultural Crops. It has been identified for the cultivation in U.P., Punjab, Bihar and Jharkhand.

Kashi Ganga: This is an early variety derived from the cross IC-92465 x DVBG-151. Fruits are light green, length 30 cm, diameter 7 cm, fruit weight 800-900 g and yield 480-550 q/ha. It is tolerant to anthracnose and suitable for rainy and summer season cultivation. This has been identified by UP State Horticultural Seed Sub Committee and notified during the XII meeting of Central Sub Committee on Crop Standard Notification and Release of Varieties for Horticultural Crops for the cultivation in U.P., Punjab, and Jharkhand.

Pusa Summer Prolific Long: It is developed at IARI, New Delhi. It is selection from local germplasm.

Sowing time	:	Suitable for both summer and *kharif* seasons
Days to first fruit harvest	:	60 days after sowing
Fruit colour	:	Green
Fruit size	:	Long (40-45 cm), neck slightly curved
Yield	:	250-275 q/ha

Pusa Summer Prolific Round: It is eveloped at IARI, New Delhi. It is local selection from germplasm.

Sowing time	:	Suitable for both summer and *kharif* seasons
Plant habit	:	Good vegetative growth
Days to first picking	:	60 days after sowing
Fruit colour	:	Green
Fruit shape	:	Round
Yield	:	250 q/ha

Pusa Naveen: It is eveloped at IARI, New Delhi. It is selection from germplasm.

Plant habit	:	Medium vegetative growth
Days to first picking	:	55 days after sowing
Fruit colour	:	Green
Fruit shape	:	Straight, nearly cylindrical
Fruit size	:	Long (30-40 cm)
Fruit weight	:	800 g
Yield	:	325 q/ha
Miscellaneous	:	Suitable for packing for long distance transportation

Pusa Sandesh: It is developed at IARI, New Delhi. It is selection from germplasm.

Sowing time	: Suitable for both summer and *kharif* seasons
Plant habit	: Medium vegetative growth
Days to first picking	: 50-60 days after sowing
Fruit colour	: Green
Fruit shape	: Round, deep oblate
Fruit weight	: 600 g
Yield	: 320 q/ha

Arka Bahar: It is developed at IIHR, Bangalore. It is pure line selection from local collection (IIHR-20) from Karnataka.

Year of release	: Released by SVRC in 1984
Area of adaptation	: Karnataka
Fruit colour	: Attractive light green with shinning skin
Fruit shape	: Straight, devoid of crook neck
Fruit size	: Medium long
Fruit weight	: 1 kg
Duration of the crop	: 120 days
Resistance	: Tolerant to blossom end rot
Yield	: 40 to 45 t/ha
Miscellaneous	: Very good cooking and keeping qualities

Kalyanpur Long Green: It is developed at Vegetable Research Station, CUAUA&T, Kalyanpur, Kanur, Uttar Pradesh, India

Year of release	: 1982
Area of adaptation	: Uttar Pradesh
Growth habit	: Vines are vigorous and long
Fruit characteristics	: Long, green, tapering, blossom end pointed
Duration of the crop	: 125 days
Resistance	: Moderately resistant to mosaic
Yield	: 25 to 30 t/ha

Azad Harit: It is developed at CSAUA&T, Vegetable Research Station, Kalyanpur, Kanpur.

Year of release	: 1989
Area of adaptation	: Uttar Pradesh
Growth habit	: Long creeper
Fruit characteristics	: Medium long, fruit weight is 1.0 kg
Duration of the crop	: 124 days
Resistance	: Resistant to mosaic virus
Yield	: 30 to 35 t/ha

Azad Nutan: It is developed at CSAUA&T, Vegetable Research Station, Kalyanpur, Kanpur.

Year of release	:	1998
Area of adaptation	:	Uttar Pradesh
Growth habit	:	Long creeper
Fruit characteristics	:	Long, green, fruit weight is 1.0 kg
Yield	:	35 to 40 t/ha

Samrat: It is developed at MPKV, Rahuri. It is selection from local germplasm collected from local village Dahanu.

Year of release	:	1992
Fruit colour	:	Green with dense pubiscence
Fruit shape	:	Cylindrical
Fruit size	:	Long 30-40 cm
Duration of the crop	:	180-200 days
Yield	:	40 t/ha
Miscellaneous	:	good for box packing and keeping qualities

Punjab Komal: It is developed at PAU, Ludhiana.

Year of release	:	1988
Maturity	:	Early
Days to first harvest	:	70 days after sowing
Fruit shape	:	Oblong
Fruit size	:	Medium
Fruit colour	:	Light green with pubescence
Fruits per vine	:	10-12
Yield	:	40 t/ha
Miscellaneous	:	Tolerant to cucumber mosaic virus

Punjab Long: It is developed at PAU, Ludhiana.

Year of release	:	1997
Fruit shape	:	Cylindrical
Fruit colour	:	Light green
Yield	:	18 t/acre
Miscellaneous	:	Suitable for packing & long distanced marketing

Pusa Meghdoot (F1 hybrid): It is developed at IARI, New Delhi. It is Pusa Summer Prolific Long x Selection-2.

Sowing time	:	Suitable for both summer and rainy seasons
Days to 50% flowering	:	Early
Fruit characteristics	:	Green, long and tender

Yield	: High yielding giving 15% increase in early yield and 74% increase in total yield over Pusa Summer Prolific Long.

Pusa Manjari (F1 hybrid): It is developed at IARI, New Delhi. It is Pusa Summer Prolific Round x Selection-11.

Sowing time	: Suitable for both summer and rainy seasons
Days to 50% flowering	: Early
Fruit characteristics	: Green, round, tender and attractive
Yield	: Gives nearly 48% total higher yield over Pusa Summer Prolific Round

Pusa Hybrid-3 (F1 hybrid): It is developed at IARI, New Delhi.

Area of adaptation	: Northern plains
Sowing time	: Suitable for both summer and rainy seasons
Days to first picking	: 50-55 days after sowing
Fruit colour	: Green
Fruit shape	: Slightly club shaped without neck
Fruit weight	: 1 kg
Yield	: 42.5 t/ha in summer and 47 t/ha in rainy season
Miscellaneous	: Suitable for easy packaging in cardboard boxes for distant marketing

Pant Sankar Lauki-1 (F1 hybrid): It is developed at GBPUAT, Vegetable Research Centre, Pantnagar. It is hybrid between PBOG-22 x PBOG-40.

Year of release	: By CVRC & UPSVRC in 1999 simultaneously
Area of adaptation	: Suitable for planting in plains as well as in hills
Sowing time	: Suitable for both summer and rainy seasons
Days to first picking	: 60 days after sowing
Fruit characteristics	: Green, intermediate long and somewhat cylindrical (about 35 cm long)
Vine length	: About 5.5 meter
Yield	: 40 t/ha

Pant Sankar Lauki-2 (F1 hybrid): It is developed at GBPUAT, Vegetable Research Centre, Pantnagar. It is hybrid between PBOG-22 x PBOG-61.

Year of release	: 2001 by UPSVRC
Area of adaptation	: Suitable for planting in plains as well as in hills
Sowing time	: It can be sown from March to July in plains and April to May in the hills
Days to first picking	: 65 days after sowing

Fruit characteristics	:	About 40 cm long, club shaped with smooth green colour
Yield	:	40 to 45 t/ha

Azad Sankar-1 (F1 hybrid): It is developed at CSAUAT, Kanpur.

Year of release	:	2001
Area of adaptation	:	Uttar Pradesh
Growth habit	:	Long creeper
Fruit characteristics	:	Medium long, cylindrical, attractive, smooth and sweet
Yield	:	50 to 55 t/ha

Varad (F1 hybrid): It is developed at Mahyco Vegetable Seeds Ltd., Jalna.

Area of adaptation	:	Recommended for cultivation in zone V
Days to first harvest	:	60-65 DAS
Fruit colour	:	Bright green
Fruit shape	:	Cylindrical
Fruit weight	:	600-750g
Fruit length	:	40-45 cm
Yield	:	60 to 65 t/ha
Miscellaneous	:	Good keeping quality

CO-1: It is developed at TNAU, Coimbatore. It is selection from local germplasm.

Fruit colour	:	Light green
Fruit shape	:	Round at base with prominent bottle neck at top
Fruit weight	:	2.02 kg
Yield	:	36 t/ha
Duration	:	135 days

Narendra Reshmi: It is developed at NDUAT, Faizabad. It is selection from the local germplasm.

Area of adaptation	:	Punjab, UP, Bihar
Days to first harvest	:	60 DAS
Fruit colour	:	Green
Fruit shape	:	Bottle shaped, shallow neck
Fruit weight	:	1.0 kg
Yield	:	30 t/ha
Miscellaneous	:	Moderately tolerant to red pumpkin beetle, PM & DM

Punjab Round: It is developed at PAU, Ludhiana.

Year of release : 1997
Fruit shape : Round
Fruit colour : Light green
Yield : 38.75 t/ha

Rajendra Chamatkar: It is developed at NDUAT, Faizabad. It is indigenous local collection from Patna region of Bihar.

Fruit shape : Long, symmetrical
Fruit weight : Medium sized, 1.35 kg

It is Early maturing, deep green foliage, fruit sets at 8th node.

Other varieties: KBG-93, Pocha Long White, Jagtial Long, Doodhi Long White, All Seasons Long, Hot Season Long White, White Surat.

12

Squashes and Pumpkins

Squashes and Pumpkin (*Cucurbita*, spp, 2n = 2x = 40) are the words quite often used interchangeably. However, the term squash is more commonly used for C. *pepo* which is consumed as an immature fruit. The term pumpkin is normally applied to the edible fruit of any species of *Cucurbita* utilized when ripe as a table vegetable or in pies.

Origin: The genus *Cucurbita* is native to America. The center of diversity is the tropics near the Mexico- Gauatamla border. The archaeology evidences indicate that these were widely cultivated in South Western United States, Mexico, and northern South America in Pre-Colombian times i.e. prior to 1492 A.D.

C. *pepo* : Summer squash : Some varieties belong to winter squash, pumpkin and marrows. It has bushy habit, peduncle hard, sharply angular, grooved, leaves broad, triangular in outline, usually with deep lobes little or no expansion of peduncle at fruit attachment. In North America it is commonly found.

C. *mixta* : Squash, cymlins, cushaws: It is used as summer and winter squash, leaves shallowly lobed, hard corky peduncle. This is also commonly found in North America.

C. *moschata* : Winter or baking squash, field pumpkin: It has vine growth habit, foliage soft hairy, leaves shallowly lobed, peduncle hard, smoothly grooved, flared at fruit attachments, popular in Central and Northern portion of South America.

C. *maxima*: Winter or baking squash: Large pumpkins belong to this species. This is regarded with high quality, soft skinned winter squash with vigorous vines and sometimes extremely large fruits, stem is soft and round, leaves rounded not lobed, peduncle soft enlarge by soft cork, fruits globular, oblong or cylindrical. These are popular in South America.

C. *ficifolia* : Plants have leaves with fig leaf shape, grooved plants, perennial, seeds are black, long vines and foliage prickly, leaves lobed fig like, peduncle hard and small, slightly expanded at the fruit attachment.

Peduncle : Key character for species identification

C. *pepo* : Deeply furrowed and 5 to 8 ridged.

C. *moschata*: 5 ridged and flared at the point of fruit attachment.

C. *maxima* : Cylindrical or claviform but never had prominent ridges.

C. *mixta* : Five angled, rounded but not flared at fruit attachment.

Gene symbols and traits controlled (Robinson *et al.*)

a → androecious

B → bicolour fruit: yellow fruit pigment before mature.

Bi → bitter fruit

bl → blue fruit colour incompletely recessive to green

Bu → bush habit, short internodes dominant to vine habit

C → coloured fruit, green fruit ; epistatic to r

r → recessive white fruit colour hypostatic to C.

Rd → red external fruit colour dominant to green

cu → cucurbitacin

D → dark green stem dominant to light green stem

Di → disc fruit shape dominant to spherical

Hr → hard rind dominant to soft fruit rind.

lt → leafy tendril (tendril with laminae)

ly → light yellow corolla recessive to orange yellow

M → mottled leaves.

ms1 and ms2 → male sterility

n → naked seeds (seeds without lignified seed coat)

Pm → powdery mildew resistance (*Spherothica fuligenea*).

ro → rosette leaf

St → striped fruit

W → white fruit, dominant to green mature fruit partially epistatic to Y

Wf → white flesh colour dom to cream flesh colour

Wt → warty fruit dominant to smooth

Y → yellow fruit colour dominant to green.

ys → yellow seedling

Upper case letter indicate dominance of allele.

Objectives of Breeding

1. High fruit yield & Early fruiting
2. First pistillate flower at early node
3. High female to male ratio
4. Yellow or mottled skin fruit
5. Non ridged fruit surface.

Breeding Methods: No inbreeding depression and hence, inbreeding and selection is ideal (Whitaker, 1974)

Hybrid breeding: Heterosis is predominant in C.*pepo* and hence F1 hybrids can be better option for improvement of pumpkins.

Methods of F1 Seed Production: Manual pollination or use of insects for pollination can be followed to cross the parents. Several rows of female line and one row of male line can be planted for F1 seed production. Buds can be detected several days before anthesis. It is easy to remove the male buds as the process of emasculation.

Use of chemicals

1. Spraying of 250 ppm ethephon several times can suppress the male flowers (Robinson *et al*, 1970).
2. Spraying 600 ppm ethephon twice applications at 2 and 4 leaf stages and there will be complete male suppression during fruiting stage. (Shanon and Robinson)
3. Ethephon is commercially used for hybrid seed production in squashes (Swaroop, 1991).

Interspecific Hybridization

Among the four annual species of *Cucurbits* information is available in summer squash (Whitaker and Robison, 1986). F1 hybrids can be obtained with difficulty between any two species. Such hybrids are normally sterile. No evidence of spontaneous crossing among these species reported. Amplification of *C. maxima* X *C. moschata* was fertile and produced quality fruits. Seeds of interspecific hybrids of *C. moschata* X *C. maxima* are on sale by sakota seed company, Japan.

Methods: MM **double** → MMMM X PPPP ← **double** PP

↓

F1 MMPP X MM Back cross

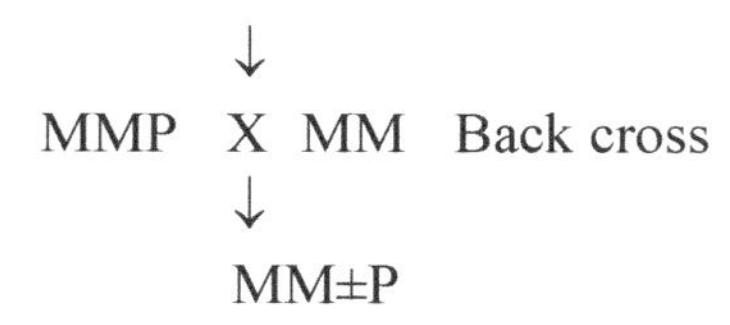

Where M: *C. moschata* , P: *C. palmata*

This is the promising procedure for transfer of desirable genes from wild (C. *palmata*) to cultivated species (C. *moschata*).

Rapid propagation of *C.pepo*: Plants have been regenerated from hypocotyls and cotyledon explants of pumpkin as reviewed by Pink and walkey (1984). This can be useful in breeding activities.

Varieties Developed

Punjab Chappan Kaddu

Selection from segregating material (*C. pepo*) bushy, 800g fruits, ten fruits/ plant, 60 days from sowing (early). It is developed at PAU, Ludhiana. Inbred selection from the segregating local material of Punjab.

Year of release	: 1982
Maturity	: Early
Fruit shape	: Disc shaped with flat stem
Fruit colour	: Green
Yield	: 20 to 25 t/ha
Miscellaneous	: It has a predominant female tendency, field resistant to downy mildew and tolerant to CMV, powdery mildew and red pumpkin beetle.

Patty Pan (*C. pepo*): Introduction from USA (1972), disc shaped fruits, chalky white, tender and very attractive at edible stage. Yield of 200q/ha in 85-90 days. Introduction from USA and recommended by IIHR, Bangalore, during 1972.

Growth habit	: Bush type
Fruit colour	: Attractive chalky white
Fruit shape	: Disc shaped
Fruit size	: Medium
Yield	: 20 t/ha in 85-90 days after sowing (short duration variety)

Early Yellow Prolific (*C. pepo*): Introduced by IARI, Regional Station, Katrain

Growth habit	: Bush type Days to first fruit harvest: Early
Fruit colour	: Skin light yellow, turning orange-yellow on maturity

Fruit shape : Warted, tapering towards stem, flesh tender

Australian Green (*C. pepo*) : Introduced by IARI Regional Station, Katrain

Growth habit : Bush type

Days to first fruit harvest : Very early

Fruit colour : Dark green with longitudinal stripes of white colour all over

Fruit size : Long (25-30 cm)

Number of fruits per plant : 15-20

Yield : 15 to 16.5 t/ha

Pusa Alankar: Developed at IARI Regional Station, Katrain. F1 hybrid between EC-27050 X Sel.-1 PI-8 (a derivative from cross Chappan Kaddu X Early Yellow Prolific)

Days to first fruit harvest : Early (45-50 days)

Fruit colour : Uniform dark green fruits with light colored stripes

Fruit shape : Slightly tapering towards the stem, flesh tender

Fruit size : Long (25-30 cm)

Fruit maturity : Mature in 45-50 days after flowering

Yield : 20 to 30 t/ha

Arka Chandan (C. *moschata*): Developed at IIHR, Bangalore. A pure line selection from Rajasthan collection (IIHR-105)

Year of identification : 1984 by SVRC and in 1987 by AICRP

Area of adaptation : H.P. and Karnataka

Pays to maturity : 125 days

Fruit colour : Light brown with creamy patches at maturity

Fruit shape : Round with flat blossom end

Fruit size : Medium

Fruit weight : 2-3 kg TSS: 8-10 %

Yield : 30 t/ha in 120 days

Miscellaneous : Rich in carotene (3331 IU/100 g), pleasant aroma, cooking and keeping qualities are good.

CM -14 (C. *moschata*): KAU, vallanikara 1987 fruit flat round, 6 kg, green in colour, shallow furrowed surface flesh thickness 4.30 cm, 15 kg/ plant yield.

Pusa Biswas (C. *moschata*): Selection from local collection SM-107 at IARI New Delhi (1987). Fruits light brown with thick, golden yellow flesh, fruits round (5kg) 200 q/ha in 120 days.

Developed at IARI, New Delhi. Local selection of line SM-107

Year of identification	: 1987 for zones V and VII
Sowing time	: Spring-summer
Growth habit	: Vigorous vegetative growth
Fruit colour	: Light brown with thick, golden yellow flesh
Fruit shape	: Spherical, flesh thick
Fruit size	: Medium
Fruit weight	: 5 kg
Days to maturity	: 120-125 days
Yield	: 35-40 t/ha

Arka Surymukhi (C. *maxima*): Improved over local collection (IIHR-79) from Mangalore (1987). Fruits small, round with flat ends, deep orange with creamy white streaks on the rind, 1 to 1.5 kg/fruit.flesh orange yellow. Keeping and transport quality good. It is resistant to fruit fly. Yields 30 t/ha in 100 days.

It is an improvement over a local collection (IIHR 79 from Mangalore

Area of adaptation	: South India
Year of identification	: 1984 by SVRC and in1987 by AICRP
Days to maturity	: 100 days Flesh colour: orange yellow
Fruit colour	: Deep orange with creamy white streaks on the rind
Fruit shape	: Round with flat ends
Fruit size	: Small
Fruit weight	: 1-1.5 kg
Resistance	: Against fruit fly
Yield	: 30-35 t/ha in 100days
Miscellaneous	: Keeping and transport qualities are good.

Narendra Amrit

It is *C. moschata* developed and released at NDUAT, Kumarganj, Faizabad. Fruits are flat- round with uniform cream colour skin. It is rich in vitamin A and has excellent flavour.

CO-1 (*C. moschata*): Developed at TNAU, Coimbatore. Selection from local type form Thudiyatur (Comibatore)

Sowing time	:	June-July to November-December and Dec.-Jan. to May-June
Growth habit	:	Plant is vigorous spreading upto 1200 cm
Fruit shape	:	Globular, flattened at the base
Flesh thickness	:	About 4-5 cm
Fruit colour	:	Immature fruits are dark-green and turns to brownish-orange after full maturity
Fruit length	:	34 cm
Fruit girth	:	26 cm
Fruit weight	:	7-8 kg
Fruit per plant	:	7-9
Days to maturity	:	175-180 days
Seediness	:	Low
Yield	:	30 t/ha
Miscellaneous	:	Good cooking quality, stem often roots at the nodes and has 8-10 branches. The peduncle is angled

CO-2 (*C. moschata*): Developed at TNAU, Coimbatore. A selection from local type

Sowing time	:	December to April, June to October
Pubscence petiole	:	Green, length 10-12 cm, girth 0.2 - 0.3 cm
Leaf size	:	14.8 x 72 cm
Days to maturity	:	135 days
Days to 50% flowering	:	50 days after sowing
Days to first fruit harvest	:	55-60 days
Fruit colour	:	Green
Fruits per plant	:	10-12
Fruit length	:	30-35 cm
Fruit girth	:	6-8 cm
Fruit weight	:	1.5-2.0 kg
Number of seeds per fruit	:	24-30
Resistance	:	Moderately resistant to insect-pests and drought
Yield	:	23-25 t/ha

Kashi Harit

This variety is derived from the cross between NDPK-24 x PKM through pedigree selection. Vines are short, leaves dark green with white spots. Fruits are green, spherical, weight 2.5-3.0 kg at green stage; yield of 300-350 q/ha in 65 days of crop duration. This has been identified by UP State Horticultural Seed Sub Committee and notified during the XII meeting of Central Sub Committee on Crop Standard Notification and Release of Varieties for Horticultural Crops for the cultivation in U.P., Punjab, and Jharkhand.

Pusa Vikas: Developed at IARI, New Delhi.

Sowing time	: Highly suitable for cultivation in spring - summer season in North India
Fruit shape	: Flattish round
Fruit size	: Small
Flesh colour	: Yellow
Fruit weight	: 2 kg
Vine length	: Semi-dwarf to dwarf (2-2.5 m long)
Yield	: 30 t/ha
Miscellaneous	: Rich in Vitamin A

Ambili (CM-14): Developed at KAU, Kerala. Selection from a local collection

Flesh thickness	: 4 cm
Year of identification	: 1987 for zone VI
Growth habit	: Spreading type
Fruit shape	: Flat-round
Fruit weight	: 5-6 kg
Fruit colour	: Green with shallow furrowed surface and turns tan coloured at maturity
Leaves characteristics	: Leaves are have white spots on the upper surface of the lamina.
Yield	: 34 t/ha and matures in 130 days

Suvarna: Developed at KAU, Kerala. Developed through single plant selection

Year of release	: 1998 by State Seed Sub-Committee
Area of adaptation	: Central zone of Kerala
Growth habit	: Trailing
Fruit shape	: Fruits flat round and green with white patches and spots at immature stage, turning yellowish brown at mature stage and flesh orange coloured
Fruit length	: 13-14 cm

Flesh thickness	:	5 cm
Fruit weight	:	3.5 kg
Crop duration	:	3-4 months
Yield	:	37 t/ha
Miscellaneous	:	Mosaic incidence can be avoided by seed sowing during September-October

Azad Pumpkin-1: Developed at CSAUAT, Kanpur.

Year of release	:	2001
Sowing	:	Suitable for both summer and rainy seasons
Fruit shape	:	Spherical
Fruit size	:	Medium
Flesh colour	:	Green, broken white pattern
Yield	:	45-50 t/ha

Pusa Hybrid -1 (F1 hybrid): Developed at IARI New Delhi.

Sowing	:	Suitable for both summer and rainy seasons
Fruit shape	:	Flattish round
Fruit size	:	Medium
Flesh colour	:	Golden yellow
Fruit weight	:	4.75 kg per fruit
Yield	:	520 q/ha
Miscellaneous	:	25% higher yield than Pusa Biswas

Saras: Developed at KAU, Kerala

Fruit size	:	Medium size.
Fruit shape	:	Elongated
Fruit weight	:	2.7 kg
Flesh colour	:	Orange
Yield	:	39 t/ha

Sooraj: Developed at KAU, Kerala

Fruit size	:	Medium size
Fruit shape	:	Round
Fruit weight	:	3.32 kg
Flesh colour	:	Orange
Yield	:	35 t/ha

Solan Badami: Developed at Dr. Y.S. Parmar UHF, Solan

Fruit size	:	Medium, 24 X 17.5 cm
Fruit shape	:	Globular with nodes
Fruit colour	:	Orange

Fruit weight	: 2-4 kg
Days to first fruiting	: 100-120 days
Yield	: 225 t/ha
Seed yield	: 4-6 t/ha

Other varieties : NDPK-224, CM-346.

13

Legumes-Garden Pea

Keys to Genera

Flowers zygomorphic, stamens definite, corolla papillonacious, petals imbricate: 1I, 1O, 3 IO

Group I: Herbs or shrubs, leaves imparipinnate, leaf lets entire, pods usually dehiscent, or if indehiscent usually small, a few seeded. *Cyamopsis*

Group II: Herbs low or climbing, leaves paripinnate, leaflets ending in a tendril or bristle, style bearded, pods dehiscent.

Sub group 1: Staminal tube oblique at mouth, pod compressed. *Vicia*

Sub group 2: Staminal tube truncate (cutting off) at mouth, pods torpid. *Pisum*

Group III: Climbing or prostrate and rarely erect

Sub group 1: style bearded, stamen diadelphous

Sub-sub group: stigma terminal: *Dolichos*

Sub-sub group: Stigma oblique: Keel spiral : *Phaseolus*

Keel not spiral: *Vigna*

Sub group 2: Style not bearded, stamen monodelphous: *Canavalia*

Group IV: Trees or shrubs, some times climbing, pods indehiscent: *Psophocarpus*

Garden Pea

It is cool season crop cultivated throughout the world. It is extensively grown in temperate zone and is restricted to cooler altitudes in tropics and winter season in sub tropics. It is rich source of proteins (25.6%), amino acids and sugars. It goes well with other vegetables in vegetable dish.

Family *leguminosae*, sub family: *papilionaceae.* 2n=14

Botanical Varieties

Pisum sativum L. Var. *hortense* is garden pea.

Pisum sativum Var. *arvense*. Field pea.

Pisum sativum Var. *macrocarpum* is edible podded pea.

Pisum sativum Var. *elatius* is wild form.

Pisum sativum Var. *sysiacum* is also wild form.

Origin: yet it is uncertain. It is considered to be originated in Ethiopia from where it spread to Mediterranean region, Central Europe, Near East ant to rest of world. *Pisum* genus has been assigned to Mediterranean and African centers of diversity. Near eastern centre is considered to be secondary centre of diversity.

Genetics: Gregor Mendel who had strong background in plant breeding and mathematics. By using pea plants, found indirect but observable evidence of how parents transmit genes to offspring. Mendel was born in1822 and was Austrian monk and studied at the University of Vienna.

Genetic studies: Heritability for days to flowering (Bhagmal, 1969), plant height (Singh *et al.*, 1973), pods per plant (Tikka and Ashwa, 1977) and for 100 seed weight were reported as high. Heritability for days to flowering (Nandapuri *et al.,* 1977), days to maturity and yield per plant were reported as low. High degree of PCV and GCV were reported for days to 50 per cent flowering, yield, number of pods, number of seeds and for number of clusters per plant (Kallo *et al.*, 1976). Genetic advance (indicator of effectiveness of phenotypic selection) was found high for number of pods per plant, plant height and for 100 seed weight (Nandapuri *et al.*, 1973). Genetic Advance was low and heritability was high for seed yield (Karannae and Singh, 1974).

Correlation studies: Pod yield per plant is positively correlated with number of days to flowering and maturity, number of seeds per plant and also with 100 seed weight. Hundred seed weight is negatively correlated with seed yield.

Path coefficient analysis: Number of pods per plant and 100 seed weight have maximum direct effect on yield (Wankanker *et al.*, 1974). Number of seeds per plant, 100 seed weight and number of days to maturity have high direct effect on yield.

Selection index: Selection index is worked out by giving maximum weightage to traits like pods per plant and seeds per plant to achieve maximum yield (Narasinghani and Kashyap, 1979).

Stability: Narasinghani and Rao (1978) obtained significant genotypic and environmental interactions for days to maturity, plant height, number of pods, seed number and seed index. Subsequently Singh *et al.* (1982) classified commercial pea cultivars as suitable for good and poor environment.

Gene action: Significant additive genetic variance was observed for seed yield per plant. Positive epiststic gene action was observed for number of pods and seeds per pod. Days to flowering is controlled by monogenic action, non additive gene action, partial dominance and over dominance as reported by different workers using different genetic stock.

Objectives

1. High yield
2. Early
3. Wrinkle seeds
4. Resistant to PM
5. Resistant to rust
6. More seeds per pod

Breeding Methods: Introduction and hybridization commonly followed.

1. **Introduction:** Arkel is an introduction from England and is early with wrinkled seeds. Bonneville is an introduction from USA and a mid maturity variety with wrinkled seeds.
2. **Back cross and pure line selection:** Arka Ajit (FC-1) is developed through backcross and pure line selection involving parents Bonneville, IIHR 209 and Freezer 656. It is resistant to powdery mildew.
3. **Heterosis breeding:** In general genetically diverse parents often produce maximum heterosis and there is a better chance of isolating transgressive segregants, but in peas geographical and genetic diversity were not related. Heterosis and inbreeding studies confirmed non significant relation between genetic divergence and hybrid performance. High degree of heterosis noted for days to flowering and seeds per pod (Singh et al, 1977). Negative heterosis was reported for 100 seed weight by Singh and Singh (1970) while, Saxena (1977) reported positive heterosis for same trait. F1 seed production system is not economically feasible.
4. **Hybridization and selection:** JM 3, JM 4, JM 1, JM 2 and PM 2 were derived by hybridization and selection. Jawahar Matar 3 was derived from T19 x Early Badger. It has wrinkle seeds and is early type. Jawahar Matar 4 was derived from T19 x Little Marvel. It has wrinkle seeds and is early type PM2 was derived from Early badger x IP3 and it is early and has wrinkle seeds.
5. **Mutation breeding:** Following mutants have been identified.
   1. **Afila :** Leaflets are converted into tendril and is governed by single recessive gene.
   2. **Acacia :** Tendrils are converted into leaflets and is governed by single recessive gene.

3. Pleiofila : Leaflets sub divided and is governed by double recessive gene.

Acacia x Afila
aaBB ↓ AAbb
F1 normal
AaBb
↓ selfing

F_2 9 normal , 3 Acacia & 3 Afila & 1 : Pleiofila
A-B-, aaB-, A-bb, aabb

Early flowering mutants identified are 46C and JP829. In these 4th or 6th node flower appeared, where as in Arkel 7th or 8th node flower appear and in Boneville 12th or 13th node flowering observed.

Fasciated mutants : Some of the faxciated mutants identified are R 701, R 710, JP 625 and JP 67. These can be used as markers. These are characterized by flower clusters at the top and not distributed along the stem and apical part of stem is band like and broadened.

Resistance Breeding

1. Powdery mildew

It is caused by *Erysiphe polygoni* and is a serious disease of North, NE, Central & other parts of India and it reduces pod number. Resistant variety JM5 was developed. Donor lines for resistance to powdery mildew are 6588, 6587, P388 etc. Powdery mildew resistance is governed by single recessive gene as reported by Narasinghani (1979); Narasinghani and Bedi, (1990).

Varieties released with powdery mildew resistance are

1. Jawahar Peas 83 (JP 83) is a mid season variety, developed from the cross (Arkel x JP 829) X (46C x JP 501). Plant height is 50 cm and each fruit contain 8 seeds and gives 120 to 130 q/ha yield.
2. Jawahar Peas 71 (JP71): It is obtained by crossing (Arkel x JP 829) X (JP 501x JM1). It is mid season variety, plant height is 50 cm, contain 6 to 7 ovules. It gives 120 q/ha yield and seeds are wrinkled and green. Hundred seeds weight is 17g.
3. Jawahar Peas 4 (JP4 & now JM6): It is derived from the cross Local yellow Batri x (6588 x 46C). It contains 5 to 6 ovules and gives 90q/ha yield. Hundred seed weight is 15g and suited for hilly area.
4. PRS4: It is a medium duration cultivar and yield is 120 q/ha and contains 6 to 7 ovules.

2. Rust resistance

Caused by *Uromyces viciiae fabae*. Powdery mildew and rust compete each other to occupy space on plant surface. Both are obligate parasites. Once powdery mildew occupies then there is no place for rust and vice versa. Hence powdery mildew resistant varieties are highly susceptible to rust. Chemical control is not effective and it is costly. Resistant lines identified are JP Batri Brown 3 and JP Batri Brown 4. These are resistant to rust and highly susceptible to PM. Further need to incorporate both these resistant genes (PM& Rust) into one variety. Subsequently developed JP179 which, is resistant to PM and tolerant to rust.

3. Fusarium wilt resistance

No systemic approach followed.

Immune lines: Sylvia, Bible Pod, Selection 1, T17, Kalanagani, Grey Giant, Alaska, Canner King, Kelvedoni Monarch as Reported by Sen & Majumdar (1974).

Resistance under field condition: Tall White Sugar, Early Giant, Grey Badger as reported by Utikar and Sulaiman (1976). Multi racial problems also reported. High resistance is obtained in JM2, JM1, GC 468, Selection 23-3-2.

Resistance observed in Pusa Vipasha, Kalanagani selection 2, Pl-2, selection 525, GC66, Lokar, EC 3833, Canner King and Super Alaska.

JP501 A/2: it is resistance to *Fussarium* and PM as reported by Tiwari & Narsinghani.

Genetics: Manogenic dominant resistance is reported by Tiwari & Narsinghani(1991).

Insect Pest Resistance

Leaf minor resistant: JP 179, JP 169-1, JP 747 are resistant donors. Lines LMR 4, LMR 10 and LMR 20 were free from leaf minor at Hissar.

Bruchid or pulse beetle: From pod in field to storage they are carried. JP9, JP179 were found resistant. F1 susceptible and F2 segregated.

Multiple Resistance

JP179: resistant to PM, Fusarium, Bruchids and leaf minor and tolerant to rust.

JP 9: PM and Bruchid resistance.

JP 501 A/2: Fussarium wilt, PM and Bruchid resistant.

JP Batri Brown 3 and JP Batri brown 4: resistant to rust and Bruchids.

PMR lines, JP 83, JP 71, JP 72, JP 585 and PRS 4 need to be crossed with JP 179 for further improvement.

Varieties Developed

Arkel: introduction from England and tested by IARI, suitable for northern and central India. Plants dwarf having bigger and dark green pods, sickle shaped, highly susceptible to color rot, seeds wrinkled and bold. 7-8 ovules per pod, 100 seeds weight is 17gm, Yields 50 q/ha.

PM 2: developed from cross of Early Badger and IP 3, seeds wrinkled. 6 ovules per pod, 50 q/ha yield.

Hissar Harit (PH 1): It has green seeds and has yield potentiality of 100 q/ha.

Bonneville: It is an introduction from USA. Seeds are green, wrinkled and bold, 100 seeds weight is 17gm, highly susceptible to powdery mildew, 8 ovules per pod and yield is 100 q/ha.

IP 3 (Pant Uphar): Its fruit has 7 ovules and it has 100 q/ha yield potentiality.

Kashi Nandini: It is early variety with 8 to 9 cm long pods, 8 to 9 ovules, yields 110 to 120 q/ha and tolerant to leaf minor and pod borer.

Kashi Udai: It is early variety with 8 to 9 ovules and 100 to 110 q/ha yielding potentiality.

Kashi Mukti: It is early variety, resistant to powdery mildew and possess 8 to 9 bold seeds per fruit and has yielding ability of 110 to 120q/ha.

Kashi Shakti: It is medium maturity variety with 11 to12 pods per plant and yields 140 to 160 q/ha.

Vivek Matar-8: It is released from VPKAS Almora.

Kashi Samrat (IVRP-9): Late, resistant to powdery mildew, 95 q/ha.

Palam Priya : Resistant to powdery mildew and high shelling percentage (60%).

Arka Karthik and Arka Sampoorna improved tropical type varieties were released from IIHR, Bengaluru.

Edible Podded Pea Varieties

Sylvia, Matter Ageta 6, Aparna, Oregon 523, Pershot Suit, Kharkovskii, Vica, Alaska 81, Taichung 12, Taichung 13 and Pervenets.

Other Varieties

Alaska: This is an early smooth seeded canning cultivar with bluish green seeds. Pods are borne singly, light green in colour, 7 < 1.25 cm, contain 5-6 small green seeds and shelling percentage 42 (Mac Giblivary, 1961).

Alderman: This is an excellent cultivar for home garden, shipment and freezing. Pods are borne singly, 11.25 cm long with 8 to 10 seeds.

Asauji: The selection of this variety has been made from the material collected from Amritsar and released from 1ARI, New Delhi. It is a dwarf, early, green and smooth seeded cultivar suitable for early sowing. The pods are produced singly, about 8 cm long, curved, dark green, narrow and appear round when fully developed having 7seeds/pod. The pods are ready for harvesting in about 60 days. It gives a shelling percentage of about 45.

Azad P-2 (PRS-4): It is a powdery mildew resistant variety, developed at Vegetable Research Station, CSAU & T, Kanpur, derived from the cross Bonneville *X* 6587. Pods are medium with 6-7 ovules, nearly straight, light *green*, smooth and firm. Horticultural maturity duration is of 90-95 days with 120 q/ha green pod yield. Physiological maturity duration is of 125-130 days. Seeds are wrinkled and brown seeded, suitable for cultivation in late sown condition and powdery mildew prone areas.

Early Giant: It is a late variety suited to hill of Himachal Pradesh. Plants are tall and need staking. Pods are 9-10 cm long, dark green with 9-10 bold grains. Seeds are wrinkled.

Azad P-1: It is a derivative of the cross 6416 *x* 6405 and released in 1983 from Uttar Pradesh. Pods are smooth, dark green, 8-10cm long, narrow (1.2- 1.4cm) very tightly filled, 8-10 seeds/pod and 30-40 pods/plant. Pods are slightly curved at the distal end. Mature seed is wrinkled. This variety escapes powdery mildew and rust. Seed yield is 17q/ha.

Early Badger: It is a dwarf early wrinkled-seeded cultivar evolved at Wisconsin. It is suitable for sowing in early October. It is ready for harvesting in 60 to 65 days. The yellowish green pods are borne singly, 7.5 cm long, well filled, 5-6 seeded and sweet. It is a good canning cultivar having a shelling percentage 36. It is resistant to fusarium wilt and tolerates heat and drought.

Early Supperb: It is an English dwarf cultivar with yellowish green foliage. It is an early smooth-seeded variety. The pods are borne singly, dark green and curved with 6-7 seeds.

Harbhajan (EC33866): It is an introduction and suitable for all dry areas of north India. Plants are dwarf. Pods are green and long. Seeds are small (13 g/ 100 seeds), round, wrinkled and yellow in colour. It is susceptible to powdery mildew.

Hisar Harit (PH1): The variety is semi-dwarf and early. It takes 60 days for first picking. Pods are single to double, large, green, well filled, and dimpled after drying.

Tiara Bona: It is developed through pure line selection from the local field pea *variety* and released in 1989 from Punjab. Days to first picking are 45-50 days.

Pods are 7.8 cm long, dark green and 5-6 seeds per pod with 40 per cent of shelling. Green seeds are less sweet. Mature seeds are bold, round, dimpled form both sides and green in colour. Seed yield is 20 q/ha. This variety escapes powdery mildew and pea rust.

Jawahar Matar-1 (JM-I or GC141): It is a derivative of the cross T-19 X Greater Progress. Flowers are white colour, two flowers per axil. Days to initiation of flower are 55 days. Pod setting begins from 11th node and usually 2 pods per axil. Pods are straight with bead like out growth at the lower end, green colour, medium thick pod wall and 8-9 seeds/pod. The shelling percentage is 52.rhe mature seeds are green in colour, wrinkled and containing 18.9 g/100 seeds.

Jawabar Matar-2 (JM-2 or GC 477): This is a derivative of the cross Russian 2 and Greater Progress'. It has bigger pods having 9 green bigger sweet ovules. This green pod yield is around 100 q/ha. Shells are comparatively thicker and have better keeping quality suitable for transportation. The matured seeds are green, wrinkled with 100 seed weight of 17 g.

Jawahar Matar-3 (JM-3 or Early December): Wakankar and Mahadik (1961) have bred this cultivar in Madhya Pradesh. A very early variety developed from a cross of T19' and 'Early Badge'. It is a dwarf cultivar. There are 4-5 ovules/ pod. The seeds are green, wrinkled and bold (100 seed weight 18 g). The first picking can be done in 50 days after seed sowing. The yield potential of this variety is 40q/ha green pods.

Jawahar Matar 4 (JM-4 or GC 195): It has been developed from the cross 719' *X* 'Little Marvel'. Plants are 55-60 cm tall and posses medium size green pods (7.0 cm), having 5 or 6 ovules/pod. The matured seeds are green, wrinkled and bigger in size (18 g/l00 g seed). It yields about 60 q/ha

Jawahar Matar-5 (JM-5): It is released in 1980 from Madhya Pradesh. Plants are 1.5 -2.0 meters in length with large inter nodes, greenish yellow foliage, two pods/ axil and 5-6 seeds per pod. Mature seeds are yellow and wrinkled. Seed index varies from 10-20 g/100 seeds. Seed yield is 10 q/ha. This variety is immune to powdery mildew.

Jawahar Peas 4 (JP 4 and now JM6): It is a cross of 'Local Yellow Batri' x ('6588' *X* '46c'). The variety possesses medium size pods with *5-6* bigger size green ovules. It yields around 90q/ha in plants. The seeds are yellow and medium sizes (100 seed weight I5 g). The pods are ready by October end or early November.

Jawahar Peas-83 (JP-83): It is a mid-season powdery mildew resistant garden pea developed from a double cross (Arkel x 3P-829) x (46-C x JP-50l). The pods are bigger in size curved with 8 green sweet ovules. The cultivar possesses a high yield potential of 120-130q/ha green pods.

Kanwari: It is a smooth-seeded main season variety. This is a tall-growing double- podded cultivar. Flowering takes place 65-70 days and first blossoms appear at 15-17 th node. Pods are about 8.5cm long, yellowish green and *5-6* seeded with 40 percent shelling.

Khapar Kheda: It is tall growing double podded cultivar. Flowering takes place in 65-70 days and first blossom appears at 15-16th node. Pods are 5.5 to 6cm long and 4-5 seeded with 50 per cent shelling.

Kelvedon Wonder: It is having dwarf plant. Flowering takes place 40 days after sowing and first blossom appears at 8-9th node. Pods are curved, borne singly, green, about 9 cm long and 6-seeded.

Lincoln: It is a dwarf to medium-tall, single podded cultivar. Pods are dark green, 9.5 - 10 cm long, 6-7 seeded with shelling percentage 45. It is suitable for late sowing. Good for canning.

Little Marvel: This cultivar has been bred in England from the cross Chelsea Gem x Suttons Alaska. Plants are dwarf with dark green foliage. Pods are about 8cm long, borne singly thick-skinned dark green, straight and broad containing 5-6 seeds.

Lucknow Boniya: It is early smooth-seeded cultivar of in the plains (Singh and Joshi, 1970). The pods borne singly, small, green and 4-5 seeded.

Madhu: It is derivative of the cross 6126 x Sylvia, released in 1973 in Uttar Pradesh. Pods are 12-15 cm long, light green, smooth, membrane less, straight, 2 -2.5 cm broad, 20-25 pods per plant and 5-6 seeds per pod. Mature seed smooth, round and of grey colour. This variety is susceptible to powdery mildew. Seed yield is 8-10 quintals per hectare.

Matar Ageta 6: It is an improvement over Arkel' and 'Harabona' and released in 1989 by PAU, Ludhiana. Plants are dwarf, quick growing, erect with green foliage and ready for first picking within 7 weeks after sowing. Pods are long, 12-15 in number, borne singly and in pairs, well filled and containing six grains on an average. The shelling percentage is 44.

Metor: It is a round smooth seeded early variety and introduced from England. Plants are 35-40 cm tall, dark green and flowers borne generally singly. Pods are dark green, 8.7cm long, well filled with 7 seeds, having shelling percentage 45-Pods mature in 5 8-60 days and suitable for early October sowing (Choudhury and Ramphal, 1975).

NDVP-8: This mid season variety has been developed at NDUA&T, Faizabad. It gives an average yield of 100 q/ha. It is recommended for release and cultivation in 1997 for Punjab, Uttar Pradesh and Bihar.

NDVP-1O: This mid season variety has been developed at NDUA&T, Faizabad. It gives an average yield of 117 q/ha.

NP-29: This cultivar has been developed through selection at IARI, New Delhi. It is medium-tall, double-podded cultivar with dark green foliage. It flowers in 75-80 days and first blossom appears at 14-16th node. Pods are green, straight, about 7.5 cm long and 6-7 seeded.

Pant Uphar (IP-3): It is a medium tall variety (70 cm), relatively thin stem, leaflet small in size and foliage light green. Flowers are white. Pods are medium size, 7.5cm long having 7 seeds/pod and the total yield is about 100 q/ha. This variety is susceptible to powdery mildew

Perfection New Line: It is a heavy yielding mid-season variety. It is ready for first picking in 80 to 85 days (Nath, 1976).

PM-2: This variety has been produced from a cross of Early Badger' with 'IP-3'. The plants are 50-60cm tall with green colour pods, having 6 ovules. The seeds are wrinkled. The yield of green pods is about 50q/ha.

Punjab-87 (87-1): It is a derivative of the cross Pusa-2 *x* Morrasis-55 and released from Punjab. Pod is 9.3 cm in length, 1-2 pods/axil, green ovules, sweet, mature seed bold and wrinkled, 7-6 seeds/pod and 48.3 per cent shelling.

Punjab-88 (88-2-C): It is developed by selection from the cross Pusa-2' and 'Morassis 55'. Plants are tall with medium size pods and having 7 green ovules per pod. The seeds are green, wrinkled and bold. This variety is also susceptible to powdery mildew.

Sylvia: It is an edible-podded and tall growing variety. Pods are about 8.5cm long, yellowish green and 5-6 seed/pod with a shelling percentage 40.

T-19: This has been developed by Agriculture Department of Uttar Pradesh. It is a medium-tall and double-podded variety. The plants flower in 55-60 days and first blossom appears at I2-14th node. Pods are yellowish green, slightly curved, 8.5 cm long and 6-7 seed/pod having shelling percentage 45.

Vivek-6: It is a cross between Pant Upahar and VL Matar 3 and released by the Central Variety Release Committee in 1996. It is a medium-maturing variety takes 135- 140 days and 80-85 days to maturity in mid hills and plains, respectively. The yield goes up to 123.68 q/ha green pods.

VL-3: This tine has been bred from a cross of 'Old Sugar' and Early Wrinkled Dwarf 2-2-a'. The variety has a height of about 65-70 cm and possesses light green pods of 6 cm in length having 6 green ovules. Yield potential' of this variety is about 90 q/ha green pods. It is susceptible to powdery mildew.

VL-8: It is a mid season variety, developed at VPKAS, Almora. it gives an average yield of 100 q/ha.

Jawahar Peas 71 (JP 71): It is a progeny of double cross (Arkel X JP 829) x (JP 501 x TM 1) having powdery mildew resistance belongs to mid season group. The pods are medium sized (7 cm) with 6-7 ovules. The seeds are green, wrinkled and bigger. The yield of green pods is 120 q/ha

VL Ageti Matar-7: It is an early maturing and high yielding variety and developed from VPKAS, Almora (Sridhar *et al.*, 1997). It matures in 120-125 days. The green grains are dimpled bold, very sweet with high T.S.S. (16.8 per cent). It yielded 93 and 103 q/ha green pods in hills and plains, respectively. It is free from the incidence of powdery mildew.

14

French Bean

Among beans, the most popular bean is French bean and it has wider distribution and larger gene pool. Various names are associated with cultivars viz., Snap bean, salad bean and green bean and all these names associated are with beans of vegetable types. The names haricot bean, dry bean and navy bean are applicable to pulse types. String bean, dwarf bean and pole bean applicable to growth forms. It is known as Rajmah in Hindi, Tingalawari in Kannada and Fras bean in Punjabi. Cultivars preferred for Canning, frozen and freeze-dried are having round pods. Fresh market cultivars are flat or oval shaped. Yellow pod cultivars are also grown. Generally fresh market cultivar yields higher in temperate region than in tropical region. Beans are said to be antidiabetic and good for bladder burns, cardiac, carminative, diuretic.

Origin and Distribution: *Phaseolus vulgaris* was domesticated in Central America about 7000 years ago and it is also thought to be domesticated in Brazil and Northern Islands and was introduced to Europe in 16th century. Vavilov (1951) reported that *Phaseolus vulgaris* originated in South Mexico and Central American centre of origin. The first cultivar has been selected in Peru. From Peru it spread to Europe and to Asia. Wild species *P. aboriginus* in North West Argentina is considered as the wild progenitor of French bean. Based on growth habit, crop can be classified into pole type, semi pole type and bush type. Former varieties are more string types. American origin of *P. vulgaris* is questioned by De Candolle (1584). Southern part of Mexico and Central America are considered to be primary centre of origin. While secondary centre lies in Peru-Bolivian-Ecuador of South American continent. In Europe, French bean spread rapidly in 16th and 17th centuries and reached England by 1594. From Europe, it was introduced to India during 17th century (Simmonds, 1976).

Domestication of French Bean Led To

1. Reduction in number of branches and leaves,
2. Increase in leaf size and stem diameter in all species except *P. acutifolius*,
3. Change in pod and seed size
4. Increase in flower size in *P. vulgaris* and *P. coccineus*

5. Seed number per pod changed in *P. vulgaris* and *P. acutifolius*. Maximum of nine seeds were found in wild forms, but rarely more than five seeds in most of the cultivars.
6. Permeability of seeds to water has increased and that has led to uniform germination and easy cooking (Miruda, 1974).
7. Variation in colour of testa: *Phaseolus* sps.-white, black, red and brown-various patterns of spots, fleeks and strips.
8. Alteration in pod structure: Reduction in dehiscence and fibre content.

   There are three pod textures in common bean:

   i) Parchment with very fibrous and dehisce strongly at maturity, used for dry seeds

   ii) Leathery: less dehiscent, but split readily along sutures, green pods used and for haricot production when fully matured.

   iii) Fleshy: String less pods, indehiscent and do not split readily
9. Change in response to photoperiod. Primitive types are native of tropics where many cultivars are photoperiod sensitive and responsive to short days. Subsequently spread to temperate region where selection for day neutral or tolerance to long days was observed.

Taxonomy

1. *Phaselous vulgaris*	Common, haricot, navy, French or snap bean
2. *P. coccineus*	Runner or scarlet bean
3. *P. acutifolius* Var. *latifolius*	Tepary bean
4. *P. lunatus*	Lima, Sieve, butter or Madagascar bean

Cytogenetics: Chromosome number of 2n=22 observed in *Phaseolus* genera of several species (Weinstein,1933).

Crossing technique: Best time for pollination is from 7 am to 8 am. Open one side of flower by careful removal of one half of keel and extract the anthers and smear it on stigma. Bhatnagar *et al.* (1972) method assures best fruit set where emasculated flowers left uncovered assured best fruit set and stigma is brushed with a solution obtained by smearing anthers in 1% sucrose solution with boric acid.

Genetic resources: Conservation of germplasm is done by NBPGR, New Delhi at its regional stations Phagli and Shimla.

Exotic introductions are Top crop and Contender from USA, Premier, Giant Stringless and BKW-74 from Sweden, Wade, EC 24940, EC 74958 and EC

30021 from Russia and Jampa (suited for Maharashtra) from Mexico. Pole type introduction are Kentuky Wonder and Watex (suited for Nilgiri hills).

Evans (1975) observed 5 distinct types

1. Indeterminate 17-35 internodes
2. Indeterminate and semi climbers with 14 to 30 internodes
3. Indeterminate bush with 13 to 25 internodes
4. Determinate climber with 12 to15 internodes
5. Determinate bush with 3 to 5 internodes

Inheritance and Genetic Studies

Green hypocotyl is controlled by single recessive gene and dark flower colour is dominant over white (single dominant gene). Response to photoperiod where single gene is involved (Padha, 1970). Round pod is dominant over flat, long pod is dominant over short where single gene is involved. Green pod is dominant over Waxy pod with single gene. Indeterminate growth habit is dominant over determinant growth habit with single gene. Interlocular cavity governed by single dominant gene. Red or purple stem is always associated with red flower colour. Pink stem and pink flower are associated and green stem can have all the coloured flowers. White flowered plant bears white seeds, while dark coloured ones bear all types of coloured seeds.

Growth habit	Monogenic	Indet / climber domoniant to determinate.
Pod stringness	Monogenic	Stringness dominant to stringlessness
Earliness	Monogenic	Early flower dominant to late flower
Pod shape	Polygenic	Normal round dominant over other shapes (cross section)
Colour of foliage	Digenic	Green dominant to varigated
Hard seed	Monogenic	Soft seed dominant to hard seed.

Heritability: High heritablilty was observed for pod length, fruit set and number of primary branches, seed yield per plant, number of pods per plant, green pod yield per plant, 100 seed weight, plant height, secondary branches, pod weight, inter-nodal length and days to first harvesting.

Genetic variability: Studies at RS, NBPGR, Shimla using 1751 accessions revealed with variability for plant height, number of clusters, pods per plant, pod length, seeds per pod, 100 seed weight and yield per plant.

Genetic advance: Expected GA was high for pod length, seed yield per plant, pod yield per plant, days to flowering, number of primary branches, average pod weight, number of pods per plant, plant height and number of secondary branches. High heritability and high genetic advance indicate the resemblance between genotype and phenotype and hence selection based on phenotype could be effective.

Correlation Studies and Path Analysis

Green pod yield was directly and positively correlated (significantly) with number of pods per plant (Joshi and Thomas, 1987) and 100 seed weight (Joshi and Thomas, 1987) and number of nodes bearing pods (Sharma *et al*., 1977). Number of branches, plant height, pod length, primary branches and days to maturity played indirect role in yield improvement. Significant and positive association existed between pod yield and per cent crude fibre, ascorbic acid and reducing sugar contents (Rawal *et al*., 1984).

Objectives of crop improvement

1 High yield and Earliness
2 Wide adaptability
3 Photo-intensive varieties
4 Adoption of pole type in hills
5 For fresh market and processing
6 Pod bearing above leaves – easy picking
7 Determinate habit for temperate region and indeterminate habit for tropics
8 Quality- decreased fiber and increased protein content, stringless pods
9 Fruits held high from the ground level
10 Rich in protein content
11 Resistance to pest and diseases, bacterial wilt, PM, rust anthracnose and common bean mosaic, root knot nematode, aphids and leaf hopper, wilt due to stem fly.
12 Development of varieties of high fertilizer responsive.
13 Varieties tolerant to frost, heat and herbicide
14 Dry shelled beans with red colour, bold grain and more shelling percentage (non-vegetable type)

For yield improvement selection for yield and other contributing characters is important

- Total green pod yield
- Number of pods per plant
- Number of pod clusters per plant
- Long tender pods

- Flat pods – increased seed yield.
- Round to semi-round fibreless pods – vegetable types.
- Growth habit – pole types yield more than bush varieties
- More number of primary branches indicates vigorous nature of plant and is important character contributes to pod or seed yield.
- Day neutral type nothing to worry about date of sowing.
- Anthracnose susceptibility decreases yield
- Resistant varieties: SVM-1, Pusa Parvati, Lakshmi.

Centers Working on Crop Improvement of French bean are

- NBPGR – RRS, i) Phagli, ii) Simla, iii) Akola, iv) Shillong, v) Trichur
- Department of Vegetable Crops, UHF, Solan
- Division of Vegetable Crops, IIHR, Bangalore
- IARI, RRS, Katrain GBPUA&T, Pantnagar
- VPKAS (ICAR), Almora, Uttaranchal, and many other institute and universities.

Methods of Breeding

1. **Introduction and selection**: Bountiful, contender, Jampa, Premier, Giant Stringless and Kentuky Wonder were exotic introduction.

   Top Crop, Kentuky Wonder, Mexico, Watex and Contender from USA. Premier, Giant Stringless and BKW-74 from Sweden EC 29490, EC 24958 AND EC 30021 and Wade from Russia (Thomas *et al*., 1983).
2. **Pure Line Selection:** IIHR: i) Sel-5- purple pods for rainy season finally not released. ii) Sel-9- Arka Komal PLS from IIHR-60 Collection from Australia, It is erect, bush type with flat pods. IIHR-220 (Arka Bold), PLS from Hungary Collection, Resistant to Rust and Bacterial blight
3. **Pedigree method**: IIHR-909 (Arka Suvidha) Pedigree selection from Blue Crop x Contender bushy plants, photosensitive. Single cross, double cross or back cross are initially made but later pedigree selection can be followed.
4. **Hybaridization and Selection:** Lakshmi. Inter varietal cross: Premier x Local (pole type).
5. **Backcross breeding**: Arka Bold x Arka Komal were crossed and back cross breeding was followed and derived high yielding lines possessing resistance to bacterial blight and rust.

6. **SSD:** Suggested for improvement.
7. **Interspecific hybridization**: SVM-1: SVM-1 is developed from inter-specific hybridisation. *P. vulgaris* (Contender) x P. *multiflorus* (PLB 257)

Species crosses	Compatibility	Utility
P. vulgaris X *P. Coccineus L.*	F1 seeds viable, F1 plants fertile	Tolerant to frost and viruses
P. vulgaris X *P. flavescens*	Crosses readily	Rust resistance
P. vulgaris X *P. lunatus L.*	Fertility of F1 ranged from nil to normal	Resistance to diseases, Possess good quality parameters

8. **Heterosis** : Highly self pollinated crop and F1 seed production is difficult and hence commercially not feasible. Carol x Michelite cross resulted with more leaf area.

   Soviet hybrids: Observed heterosis for yield components.

   Heterosis reported for seed yield but inbreeding depression in F2, non-additive gene action was involved.
9. **Mutation breeding**
   1. Pusa Parvathi: Gill *et al*., (1972) derived this variety from a mutant waxy type EC 1906, (Seeds were exposed to X-rays). It is early, bushy, having green pods of good cooking quality, ready for pod picking in 45 to 50 days after sowing.
   2. Universal: Seeds of variety Granda were exposed to X-rays (300 rad) and derived this variety which is early and resistant to anthracnose.
   3. Sanilac: Seeds of variety Michelite were irradiated with X-rays and derived this variety which is bushy, early and resistant to anthracnose and common bean mosaic virus.
   4. D66-26 parent subjected to X-rays irradiation and obtained mutant which is early maturing, having long pods and high yielder than parent (Bhatt *et al.*, 1972).

   Effective dose of X-ray in French bean is 10 k-rad and dwarf mutant obtained by Subramanian, (1979). Use of chemical mutagen ethylmethane sulphonate (EMS) and methyl methane sulphonate (MMS) has given 0.3% frequency of mutations and 7 early flowering mutants were obtained by Pandey and Seth (1975).
10. **Polyploidy**: Sheopuria and Tiwari (1970) carried out ploidy breeding where *P. vulgaris* was subjected to colchicines treatments and obtained Tetraploids which were late maturing with large flowers, pods and seeds. Delayed seed germination, poor pod set and poor survival rate was observed in these

tetraploids. Biswas and Bhatacharya (1976) obtained highest number of tetraploid with 0.25% colchicin. Tetraploids had thick, dark green and deformed leaves. Autotetraploids (2n = 44) and monosomics (2n = 21) reported. Amphidiploids also tried but, no distinct advantages.

Quality breeding: Includes flavour, texture, skin sloughing, colour and biochemical composition especially protein.

High seed protein- Promising sources: Prince, Bountiful, Master Piece, French Horticulture, Triumph, Sure Crop, Giant Stringless, Green Pod And Premier (Lal and Padda, 1972). Local line from Ludhiana with light brown testa contained the highest lysine and tryptophan (Sekhon and Chopra, 1975). Rawal *et al.* (1984) identified 'Frill' variety with good quality and VL-Boni-1 was best for yield and quality. Mutation breeding helped in increasing protein and methionine.

Resistance Breeding

Source of Resistance

Powdery mildew (*Erysiphe polygony*)	:	Macarrao
Rust	:	West Ralius
Anthracnose	:	Cornell 49-242
Common bean mosaic	:	Red Mexican
Root knot nematode	:	Albama No.1
Aphids	:	Bruna
Leaf hopper	:	Contender

1. **Powdery mildew**: It is caused by *Erysiphe polygonii*. Resistance is controlled by single dominant gene (Anon., 1970). Resistant varieties are Lady Washington, Pinto (Yarnel, 1965), Contender, Long Kidney, Wade, Top Crop and Hungarian Yellow (Anon., 1970).
2. **Bean rust** (*Uromyces phaseoli*): Six races identified:1, 2, 6, 11, 12, 17. For Race 1, 2, 6 and 12 resistance is governed by dominant gene, while for race 11 and 17resistance is governed by incomplete dominant gene. Transgressive segeration was observed in hybrids inoculated with race 11 (Anon., 1970).
3. **Anthracnose** (*Colletotrichum lindemuthianum*): Extent of pods infected ranged from 24 to 59 per cent. Earlier two races were identified (Shyam and Chakraborthy, 1985). Wells Red Kidney is resistant to both the races. Other races: alpha, beta, gamma and delta (a, b, g, d). *Phaseolus aborigineus* is resistant to a, b and g races (Yarnel, 1965). Sinilac BC-6 is resistant to alpha, beta and gamma (a, b, g) races of the fungus (Singh *et al.*, 1986).

4. **Angular leaf spot** (*Isariopsis griseola* Sacc.): Crosses between PLB 257 (*P. multiflorus*) and Contender (*P. vulgaris*) attempted where F1 was susceptible and resistance is governed by single recessive gene.
5. **Bacterial Halo Blight** (*Pseudomonas phaseolicola):* Tolerance was found in PI 150414 to race 1 and 2 and is governed by single recessive gene. Resistance was observed in Red Mexican to race 1 and is governed by single dominant gene 'PPr' (Patel and Walker, 1967).
6. **Common Bean mosaic**: Several viruses infect bean and Virus 1 and 4 causes common bean mosaic, virus 2 causes yellow mosaic. Virus is transmitted by seed. Sources for resistance are US No. 5, Refugee and Idaho Refugee and these carry single dominant gene and cultivar Robust carry single recessive gene for resistance to the bean viruses (Ali, 1950). Cultivar Sinilac BC-6 carries resistance to several races of bean common mosaic virus (Singh *et al*., 1986).
7. **Root knot nematode**: Resistance is spotted in B-4174 line (Parthasarathy, 1986). Singh *et al*. (1981) isolated 31 resistant and 64 moderately resistant cultivars of *P. vulgaris* out of 303 total lines screened under field and green house.

Pest Resistance

Stem fly (*Ophiomya phaseoli*) is one of the production constraints in French bean. Lower population of stem fly was observed in Sel-2, UPF-191, SVM-1, Sel-9 and higher population was seen in VL Boni-1-1, Sel-5, Pusa Parvati, Contender, etc.

Screening of varieties resulted in identification of cultivars for some of the following traits/ adaptability.

General adaptation: Good- PLB 14-1, EC 43893 and EC 43900

Good seed yield: EC 113142, PLB 10-1, PLB 14-1 and Jawala

Vegetable types for hills: EC 57080, EC 108101

Good horticultural types (various regions, further testing recommended): BB Lake-92, Green Pak, Burly Bean, B-18

Northern plains: Selection 4-1 (Katrain)

Uttaranchal hills- free from mosaic virus and leaf spot, dwarf, string less, Top Crop.

Plains: Wade, BKW-74, EC 30021 and EC 24940

Non-irrigated areas: Kentuky Wonder and French White

Plains of Tamil Nadu: Premier, Master Pice, Stringless, Grey Pod and Bountiful.

UP Kumaon hills: Pusa Parvati, VL Bon-1.

Schematic diagram of development of IIHR 12-3 (Arka Anoop) (given by Dr T.S. Aghora, IIHR, Bengaluru)

I IHR 220	X	Arka Komal
Arka Bold (P_1)		(P_2)
(Rust and Bacterial Blight res.)		(Popular cultivated variety)

(Crossed during Kharif, 1991)

Backcrossed to Arka Bold and Arka Komal

during Kharif 1992 (B1 and B2 generated)

F_1X P_1 (kharif, 1992)

BC_1 X P_1 (rabi 1992, Selected rust resistant lines)

BC_2 to F_7 (from Kharif 1993, the breeding lines with combined resistance to rust, bacterial blight and high yield were selected and advanced up to BC_2F_7)

BC_2F_7 (during 1997, a breeding line, IIHR-220 x Arka Komal-91-17-2-1-4-12-3 with high yield and combined resistance to rust and Bacterial Blight was selected.

Replicated yield trials for 4 years from 1999 to 2003.

Total 12 Years required to develop and release the variety (Dr T. S. Aghora, IIHR, Bengaluru).

Future line of approach

- Resistance breeding for anthracnose and angular leaf spot.
- Tolerance to low and high temperature for germination and pod set.
- French bean is susceptible to low temperature for seed germination.
- French bean susceptible to high temperature for pod set.
- Genotypes tolerance to moisture stress: In hills French bean is grown as rainfed-crop.
- Breeding for yield and quality.

Varieties

Processing cultivars: Tender crop and Cascade

Pole types: SVM-1, Laksmi, Blue Lake, Phenomenal, Long Podded, Kentucky wonder.

Bush types: Bountiful, Premier, Contender, Jampa, Pencil Pod, Sure Crop, Refugee's Wax, Giant Stringless, Tendergreen, Tender Pod, Big Ben Red, Pusa Parvati, UPF-191, UPF-203, Pant Anupama, Arka Komal, VL Boni-1, VL-2, Watex and Burpee Stringless.

Phule Surekha was released during 2000 and was developed through pure line selection. It is semi determinate, photo and thermo insensitive, resistant to anthracnose, leaf crinkle, bean yellow mosaic viruses and wilt (stem fly).

Pole types

SVM-1: evolved from Inter specific hybridisation between *P. vulgaris* (var. contender) and *P. multiflorus* (var. PBL 257). It is resistant to angular leaf spot, suited for mid hill areas. It has stringless, 13 to 14 cm long pods, 8 to 10 seeds/ pod. It is vegetable as well as seed type, a dual purpose variety.

Lakshmi: Selection from cross between contender (bush) x local (Pole). It is Suited for mid hill areas. It has stringless, 13 to 14 cm long green pods with white seed colour. It yields 120 to 140 q/ha and tolerant to angular leaf spot.

Kentucky Wonder: Introduction from USA and bears 4 to 5 pods per cluster with 20 cm long, flattish, stringless pods and light brown seeds and yields 100 to 125 q/ha.

Bush types

Contender: Introduction from USA, flower pink, 50 to 55 day for picking, round, 13 to 14 cm long, stringless, slightly curved pods, seeds light brown, yields 80 to 95 q/ha and tolerant to PM and mosaic.

Premier: It is black seeded and bear 11 to 13 cm long, flattish pods in 55 to 60 day. It is adapted for late sowing and less susceptible to wilt and mosaic disease and yields 75 to 90 q/ha.

Pusa Parvati: Developed through mutation breeding at IARI, Regional station Katrain. It is early and takes 45 to 50 days and bears 15 to 18 cm long, round, stringless pods and yields 80 to 85 q/ha and is resistant to mosaic and PM.

Bountiful: It is introduced from USA and suited for September to October planting and yields 100 to 120 q/ha.

VL Boni-1: It is developed at VPKAS, Almora and is very dwarf (40 cm), suited for northern hills, flowers are white and purple tinged and bears round pods with stringless and pale green colour. It takes 45 to 60 days for maturity and yields 105 to 115 q/ha and suitable for multiple cropping.

Pant Anupama: It is early bearer and has upright foliage. Pods are round, stringless and first picking can be carried out from 55 to 65 days and yields 89 q/ha. It is resistant to angular leaf spot and moderately resistant to common mosaic virus.

Jampa: It is an introduction from Mexico. Pods are round, it is highly resistant to wilt and can withstand warmer conditions and yields 80 to 85 q/ha.

Arka Komal: It is a bush type and photo insensitive variety with pod yield of 17-18 t/ha. It is a pure line selection from IIHR-60 (Collection from Australia). Plants are erect and bushy and bear flat, green and straight stringy pods. Seeds are light brown, oblong and large. It has good transportation and cooking qualities. Seed yield is 1500 kg./ha and duration is 70 days. It yields 20 t/ha It is susceptible to Rust and Bacterial blight.

Arka Bold (IIHR-220): It is a bush type cultivar and photo insensitive and bears flat, medium long, 1.5 cm width, crisp and stringless pods. It yields 15 t/ha. in 70 days and is resistant to Rust and Bacterial blight.

Arka Suvidha: It is developed through Pedigree selection from the cross between Blue Crop and Contender. It is resistant to rust, plants are bushy and photo insensitive. It bears straight, oval, light green, fleshy, stringless and crispy pods. It yields 19 t/ha in 70 days.

Arka Anoop: It is resistant to both Rust and Bacterial blight and yields 19.78 t/ha. in 70 days. It bears crisp pods with less parchment. Plants are bushy and photo insensitive. It is derived from cross between Arka Bold (IIHR 220) and Arka Komal through modified Back Cross Pedigree method of breeding.

Arka Sharath: It has round, string less, smooth pods suitable for steamed beans. Pods are crisp, fleshy with no parchment and perfectly round on cross section. Plants are bushy and photo insensitive and it is suitable for both kharif and rabi seasons. It has high pod yield potential of 18.5 t/ha in 70 days.

Kashi Param: This variety has been developed through pure line selection. Plants are determinate and leaves are dark green. Pod is fleshy and pod length is 14.7c. It has round and dark green pods. It yields 120 to 140 q/ha. This has been released and notified during the XII meeting of Central Sub Committee on Crop Standard Notification and Release of Varieties for Horticultural Crops for the cultivation in J&K, H.P., Uttaranchal and M.P.

15

Cowpea

Southern pea / black eyed pea are the other names associate with cowpea. It is grown for vegetable and also as pulse crop and varieties are specific for use.

Classification of Cowpea

1. *Vigna unguiculata* (L.) Walp. Cv. *Sesquipedalis*: yard long bean or asparagus bean: Possess long pods and used as vegetable.
2. *Vigna unguiculata* (L.) Walp. Cv. *Cylindrica*: dual purpose, i.e. vegetable and pulse
3. *Vigna unguiculata* (L.) Walp cv. *Radiata*: Pulse type.

More emphasis is given for pulse types for crop improvement. Less attention is given to vegetable types. Pole and bush types are available in vegetable type. Cowpea improvement is from vine to bushy habit with low fiber content and low glucosides and increased protein content.

Origin and Distribution: Cowpea is in cultivation from ancient times in the tropics of old world. Country of origin is uncertain. In India, it has been known from the vedic times. Vavilov (1939) considered India as the primary centre of origin. But it is now generally agreed that Africa is centre of origin since wild forms are found in Africa and in Asia wild forms are absent (Smartt, 1990). Faris (1965) concluded that, cowpea arose from domestication of wild form *Vigna unguiculata* ssp *dekindtiana* in West Africa. But domestication is actually happened in Ethiopia and dissemination was West wards Africa, East ward across the Indian Ocean. It has spread from Africa to Asia and Europe through Egypt. Introduced to West Indies in 16th century by Spaniards and was taken to USA around 1700. Now cowpea is widely distributed throughout the tropics and subtropics.

Verdcourt (1970) considered *V. unguiculata* to comprise with five sub-species.

Wild sub-species

1. *Dekindtiana* (Harins) Verdc.
2. *Menensis* (Schweint) Verdc. : Out breeder

Cultivated sub-species

1. *Unguiculata* (L.) Walp - found throughout world where cowpea is grown
2. *Cylindrica* (L.) Verdc. - cultivated types evolved in Asia.
3. *Sesquipedalis* (L.) Verdc.- cultivated types evolved in Asia.

The sub species *dekindtiana* resembles the *unguiculata* than *menensis* and hence *dekindtiana* is progenitor of *unguiculata*. The sub species *menensis* on the most probable ultimate progenitor type i.e., *menensis evolved* into *dekindtiana* and *dekindtiana* evolved into *unguiculata*. It is believed that crop established in South West Asia by 2300 BC (Smartt, 1990) and arrived at South Europe by 300 BC and to American centre in 16 to 17th century from South Europe and West America (Smartt, 1990).

Taxonomy

1. Sub species. *unguiculata* synonyms *V. sinensis* is common cowpea which is vine type, sometimes erect, pods medium in length (20 to 30 cm), small seeds, include large number of variation in Africa and Asia.
2. Sub species *cylindrical* synonyms *V. catjang* and *V. cylindrica* is catjung bean. Plants are erect, pods small (7 to 13 cm) and erect, seeds are small (5-6 mm long) and kidney shaped.
3. Sub species *sesquipedalis* synonyms *V. Sesquipedalis. V. sinensis* var. *Sesquipedalis* is Asparagus bean or Yard Long Bean where plants trailing or climbing, pods pendent, 30-90 cm long, seeds elongated, kidney shaped 8 to 12 mm long, cultivation found in Indonesia, Philippines and Sri Lanka. These are also grown in India.

Related species are *V. luteola*, *V. nilotica* and *V. marina*.

Cowpea is a true diploid with 2n=22.

Objectives of Breeding

1. Development of high yielding and early bearing varieties.
2. Wide adaptability to suit varied agro climatic conditions
3. Development of photo-insensitive varieties for successive planting during the year.
4. Development of dwarf types to avoid staking.
5. Development of varieties with good quality- high protein, high methionine, high other essential amino acids content.
6. Long tender and string less pods for fresh consumption

7. Short tender pods for whole pod processing
8. Varieties suited for intercropping
9. Resistance to diseases: anthracnose, rust, powdery mildew and cowpea mosaic virus
10. Resistance to insects: thrips, aphids, pod borers and root knot nematodes

Genetic Resources

NBPGR has around 2554 accessions comprising primitive cultivars, old obsolete varieties and locally adapted land races. From this, 266 are from Nigeria. The exotic collectios have high degree of photo-insensitivity. Collections from India include accessions belonging to *V. unguiculata, V. cylindrical and V. sesquipedalis*. Collections from Rajasthan and Gujarat possess earliness, small pods and tolerance to viruses. Tribal tracts of Madhya Pradesh include vegetable types and from Maharashtra and Tamil Nadu include late maturing types. Pod length vary from 7 to 13 cm in *cylindrica* (HP) and 30 to 50 cm in *sesquepedalis* which were collected from Gujarat, UP and TN.

Photo insensitive varieties: Rituraj, P 460-1-1, Red Seeded, P 85-2/E9A, EC 5000 (Chandel and Singh, 1987); Sel. 24-1A, EC 38215, EC 100089 (Singh *et al*., 1976); Pusa Phalguni (Tikka *et al*., 1979).

High Green pod yield and wide adaptability source is 'Aseem'.

Sources for high Protein content: 1) *sesquepedalis* Var. Yard Long (28.75%), 2) Lalita (25.42%), 3) West Bred (25.39%) (Shivashankar *et al*.,1974).

Cytological Studies

The chromosome complement of cowpea is 2n=2x=22, where 1 short, 7 medium and 3 long chromosomes are there. Sometimes, 2n=24 also reported in *V. catjang* and *V. sinensis*. All sub-species of *V. unguiculata* are interfertile.

Roy and Richharia (1948) found difficult to cross *V. sesquipedalis* and *V. sinensis*. The cultivar CO-2 was derived from cross between *V. catjung, V. sinensis* and *V. sesquepedalis* which is high yielding and fertile.

Heritability: High estimate of heritability was reported for Yield per plant, 100 seed weight, number of pods, number of days to flowering, pod length, pod weight, cluster number, seed yield and harvest index by various workers

Genetic advance: High genetics advance is reported for yield per plant, number of branches per plant, pod length, number of pods per plant, 100 seed weight, seed yield per plant and pod weight by various workers.

Correlation studies: Pod yield is significantly associated with number of pods per plant, number of seeds per pod, pod length, number of days to 50 per cent flowering, number of clusters per plant, 100 seed weight and number of branches as reported by various workers.

Selection indices: Selection based on discriminate function of three components, *viz.*, number of pods per plant, number of grains per pod and 100 seed weight was 33% more effective. Selection should be based on plant height, seed yield, 100 seed weight and pod length. Most stable component is 100 seed weight. Hundred seed weight and number of pods per plant are better selection indices for improving the pod or seed yield.

Genotype x environment interaction is significant and hence location specific cultivars are recommended.

Inheritance

i) Growth habit: For growth habit different sources with different gene action reported where monogenic, two duplicates genes and three genes reported.

ii) Anthocyanin pigmentation: Pigmentation is dominant over non pigmentation. Purple pigmentation is controlled by two duplicate dominant genes. Pod pigmentation controlled by two complementary genes and also reported to be Pleiotrophy.

iii) Tendril: It is monogenically controlled.

iv) Leaf shape: Reported to be controlled by four genes, LS1, LS2, LS3, LS4

Hastate : Two lobes of sagittate leaves extended outward.

v) Flower colour: Petal colour controlled by single gene where violet is dominant white and intermediate is light violet indicating codominance.

vi) Photo-insensitivity: Photo insensitivity is dominant over sensitivity and is monogenically controlled.

vii) Pod number: non-additive

viii) Pod length : Partial dominance

ix) Seed colour: Black is dominant over red where one gene involved, digenic and trigenic also reported.

x) Hilum colour : Monogenic and Black is dominant over white.

xi) No. of seeds/pod : Over dominance

xii) Seed weight : Additive

xiii) Seed length : Monogenic

xiv) Seed shape : Monogenic

xv) Seed yield : Non-additive

Genetics of Characters

Trait	Genetics	Gene symbol	Description
Pod colour	Monogenic	Gp/Gp	Green dominant to white
Flower colour	Monogenic	Pf/Pf	Purple dominant to White
Standard petal colour	Monogenic	Ystp / Ystp	Yellow dominant to White
Grain length	Monogenic	Lg/Lg	Long dominant to short
Leaf shape	Oligogenic	Ls1 to Ls4	Hastate leaves dominant to rhomboid
Seed coat colour	-	-	Purple dominant to other colour
Leaf foliation	Monogenic	-	Trifoliate dominant to unifoliate
Cotyledon colour	Monogenic	-	Purple dominant to Green

Breeding Methods: Cowpea is strictly a self pollinated crop where selection, mutation and hybridization are followed for crop improvement. Flower opening time is from 7 am to 9 am and anther dehiscence takes place between 10 am and 0.45 pm. For crossing emasculation can be carried out on preceding evening and pollination by next morning. Pod set is reported to be 18.6 to 26.11% on artificial crossing.

1. **Backcross and Pure Line Selection:** Arka Garima (Sel 61-B) improved variety is derivative from cross between TU. V. 762 and *V. unguiculata* ssp. *sesquipedalis* and developed by following backcross and pure line selection.

2. **Pedigree Method :** Arka Suman (Sel-11) an early and photo insensitive variety developed through peigree selection from cross between Pusa Komal and Arka Garima. Arka Samrudhi (Selection 16) early and photo insensitive variety developed through pedigree selection from cross between Arka Garima and Pusa Komal.

3. **Heterosis Breeding:** Heterosis was observed for protein content and number of pods per plant. However, F1 hybrids seed production is difficult and hence, heterosis is not exploited.

4. **Interspecific Hybridization:** Five sub species are cross compatible. *V. luteola* is difficult to cross with cultivated species and *V. nilotic & V. marina* are not compatible with cultivated species. These species can be used as source of resistance to pest and diseases.

5. **Mutation Breeding:** VITA 25 is an erect breeding line carrying male sterility genes (ms2ms2) developed through mutation breeding. Male sterile mutant was derived from *V. sinensis* var. c*atjung*. By using EMS treatment mutants like *albino, xanthin, Chlorine* and *Striata have been developed.*

Pusa Phalguni seeds exposed to mutagenic treatments and line V20 (Asthana, 1988) developed as improved variety.

Resistance Breeding

Bacterial Blight: Bacterial blight is caused by *Xanthomonas compastris* pv. *Vignicola.*

1. Out of 105 varieties and hybrids screened under artificial conditions, 6 were resistant, 29 were tolerant (Tyagi *et al.*, 1978).
2. Out of 221 screened by Prakash and Shivashankar (1982), 16 were found resistant.
3. Of 234 lines, 166 were brown hypersensitive resistant (BHR) to race 1 and susceptible to race 2 and 3. Totally 23 were BHR to race 1 and 2, but susceptible to race 3 and 11 were hypersensitive resistant (R) to all these races (Patel, 1981).

Resistance is governed by i) single dominant gene in P309, P426 and P910, ii) single recessive gene in Iron, iii) with modifier in P910 and Iron (Singh and Patel, 1977).

Rust: caused by *Uromysis phaseolli* (organism). Resistant Source is Missisippi Crowder variety. Cytoplasm x Genic interaction reported, a single gene R is involved (Satyanarayan *et al.*, 1980).

Cowpea Mosaic: Source- EC 4219, P304, P1093, MS9804, *etc.*

Multiple disease resistant variety CO4 (Selvaraj *et al.*, 1986) is developed.

Breeding for Resistance to Insects

Aphid (*Aphis craccivora*): Resistance is dominant trait (genes AC1, AC2, Roc.). Resistant Lines identified are TVu 408 P2, TVu 46, TVu 3509, PI473 and PI476. TVu -36, TVu 801 and TV-u 3000 carry single dominant gene Rac.

Weevil: Resistance has additive, dominance and maternal components.

Bruchids and thrips: Resistance is a recessive trait and digenically inherited.

Drought: Drought tolerant lines identified are CO-1, S 488 and CO4.

Breeding Achievements

Improved Varieties: Pusa Phalguni, Pusa Barsati, Pusa Dofasali, S 203, Yard Long Bean (*V. Sesq.*), Pusa Komal (Sel. 1552), Pusa Rituraj, Philippines Early, C152, S488 and many other varieties have been developed.

Bidhan Barbati 1 (BCKV1): tolerant to mosaic virus; non-viney, high yield – developed by involving *Unguiculata* (EC 243954) and *sesquipedalis* (EC 305827). Inter sub cultivar group hybridisation followed by modified backcross – pedigree method. High protein content, compact, det, 25.2 cm pod length, 13.4 t/ha.

Bidhan Barbati 2 (BCKV2): It is derived from cross between *Biflora* (V-70) and *Sesquipedalis* (Sel. TM3) where Modified back cross-pedigree method was followed. It is Semi determinant and has yellowish green foliage with 25.8 cm long pods and has yielding ability of 15.9 t/ha.

Arka Samrudhi (Sel. 16): It has same pedigree of Arka Suman. It is F9 generation entry of pedigree Selection involving parents Pusa Komal and Arka Garima. It is photoinsensitive and takes 70 to 75 days for maturity.

Kanakamani : It is the variety released from KAU and is suited for rainy and summer season and it is suited for vegetable and pulse purpose.

'Goit' and 'Alabunch' varieties are resistant to cowpea mosaic. 'Grant' variety is resistant to fusarium wilt. 'Chinese Red' variety is resistant to phytopthora stem rot. 'Mississipi 57-1' and 'Iron' varieties are resistant to root knot nematode.

Cowpea

Swarna Harita (Pole type)

- Developed through pure line selection from IC- 28143
- Pod : Dark green, very long (50-60 cm), straight, round and fleshy having excellent cooking quality
- Matured dried seeds are light brown, elongated and kidney shaped
- Tolerant to cowpea mosaic virus and rust
- Time of sowing: February-March and June-July
- Seeds rate: 5-6 kg/ha
- Maturity : First harvest 50-55 days after sowing
- Fresh pod yield : 300-350q/ha
- Recommended for Jharkhand and Bihar

Swarna Shweta

- Developed through pure line selection from CHCP-1
- Pod : Medium long (30-35 cm), white straight, round fleshy having very good cooking quality
- Resistant to cowpea mosaic virus and rust
- Time of sowing: February-March and June-July

- Seeds rate: 5-6 kg/ha
- Maturity : First harvest 50-55 days after sowing
- Fresh pod yield : 250-300 q/ha
- Recommended for Jharkhand and Bihar

Swarna Suphala

- Developed through introduction from IC-202932
- Pod : Straight light green (30-35 cm) with bulged appearance of seeds
- Matured dried seeds are mottled brown (bicolour brown and sandalwood colour)
- Resistant to rust and cowpea mosaic virus
- Time of sowing: February-March and June-July
- Seeds rate: 5-6 kg/ha
- Maturity : First harvest 50-55 days after sowing
- Fresh pod yield : 250-300 q/ha
- Recommended for Jharkhand, Bihar, Karnataka and Kerala

IIHR-16

IIHR-16 is an early maturing variety, developed at IIHR, Banglore; derived through pedigree selection from the cross Arka Garima x Pusa Komal. Plants are erect, bushy, 70-75 cm tall, photo-insensitive. Pods are green, medium thick, medium long(15-18 cm), tender, fleshy without parchment, good cooking quality; gives an average yield of 19 t/ha in 70-75 days of crop duration. Recommended for cultivation and release in 1998 for the zones VI (Rajasthan, Gujarat, Haryana and Delhi) and VII (M.P. and Maharashtra).

Arka Garima (Sel 61-B): It is the derivative of the cross T.U.V.762 x *V. uniquiculata* sub sp. *Sesquipedalis* and developed by back cross and pure line selection. Plants are tall, vigerous, bushy, with small vines and it is photo insensitive. Leaf colour is light green. Flower colour is purple. Pods are light green, long, thick, round, fleshy and string less. It is Suitable for vegetable purpose. It is tolerant to heat, drought and low moisture stress. It takes 90 days for maturation and has pod yield potentiality of 18 t/ha.

Arka Suman (Sel.11): It is the F9 generation selected entry of pedigree selection of the cross Pusa Komal x Arka Garima. Plants are erect, bushy, photo insensitive and early with pods above the canopy. Pods are medium long, tender, fleshy crisp and without parchment and has good cooking qualities. It is resistant to rust and requires 70 to 75 days for maturity with yielding ability of 18 t/ha.

Sharika (S 108): It is a high yielding vegetable cowpea (Yard long bean) variety from KAU. It has high yielding potentiality (10.6 t/ha) and is pole type. It has long white pods with purple tip and average pod length is 49 cm.

Malika (Sel 7): It is a high yielding (9.8 t/ha) pole type vegetable Cowpea (Yard long bean) variety released from KAU. It has long light green pods without purple tip and average pod length is 43.5 cm.

Vyjayanthi (VS 21-1): It is high yielding (12.6 t/ha) pole type vegetable Cowpea (Yard long bean) variety released from KAU. It has extra long, wine red coloured pods and average pod length is 50.6 cm and average pod weight is 16.17 g.

Lola (VS 13-2): It is high yielding (20 t/ha) pole type vegetable Cowpea (Yard long bean) variety released from KAU. It has extra long, light green pods with purple tip and average pod length is 53.38 cm and average pod weight is 22 g.

Kairali (CWP 11-1): It is high yielding (7.1 t/ha), semi trailing vegetable Cowpea variety released from KAU. It has medium long, pink pods with average pod length of 22.78 cm and average pod weight of 7.08 g.

Bhagyalakshmy (VS 389): It is high yielding (6.48 t/ha) bush type vegetable Cowpea variety released from KAU. It has medium long, light green pods with average pod length of 27 cm and average pod weight of 7.13 g.

Kanakamany (PTB 1): It is high yielding (7 t/ha) semi trailing dual purpose vegetable cowpea variety released from KAU. It has medium long, dark green pods with average pod length of 17.2 cm and average pod weight of 6.2 g.

Kashi Nidhi: It is a dwarf and bush type, early, high-yielding (110-120 q/ha) variety resistant to cowpea golden mosaic virus disease. On an average, it produces 30-35 pods with 30-35 cm long, dark green, fleshy and free from parchment.

Kashi Shyamal: It is bushy variety yields 70-80 g/ha. Golden mosaic virus tolerant recommended for U.P., Punjab, and Jharkhand, notified through Central Variety Release Committee.

Kashi Gauri: This is a bush type, dwarf, photo-insensitive and early variety suitable for growing in both spring-summer and rainy seasons. Flowering starts in 35-38 days and pods get ready for harvest in 45-48 days after sowing. Pods are about 25 cm long, light green, soft, fleshy and free from parchment. The cultivar is resistant to golden mosaic virus and *Pseudocercospora cruenta*, and gives green pod yield of about 100-120 q/ ha. This has been identified by Horticultural Seed Sub-Committee and Notified during XII meeting of Central Sub-Committee on Crop Standard Notification and Release of Varieties for Horticultural Crops for the cultivation in Punjab, U.P., and Jharkhand.

Kashi Unnati: This is a photo-insensitive variety. Plants of this variety are dwarf and bushy, height 40-50 cm, branches 4-5 per plant, early flowering (30-35 days after sowing), first harvesting at 40-45 days after sowing, produces 40-45 pods per plant. Pods are 30-35 cm long, light green, soft, fleshy and free from parchment. The cultivar is resistant to golden mosaic virus and Pseudocercospora cruenta, and gives green pod yield of about 125-150 q/ ha. This has been identified by Horticultural Seed Sub-Committee and Notified during XIII meeting of Central Sub-Committee for the cultivation in Punjab, U.P., and Jharkhand.

Kashi Kanchan: This is dwarf and bush type (height 50-60 cm), photo-insensitive, early flowering (40-45 days after sowing) and early picking (50-55 days after sowing) variety suitable for growing in both spring-summer and rainy seasons. Pods are about 30-35 cm long, dark green, soft, fleshy and free from parchment. The cultivar gives green pod yield of about 150-175 q/ ha and is resistant to golden mosaic virus and Pseudocercospora cruenta. This has been released and notified during the XIII meeting of Central Sub-Committee for cultivation in U.P., Punjab, Bihar, Chhattisagarh, Orissa, A.P., M.P. and states.

Kashi Sudha: Golden mosaic virus and *Pseudocercospora cruenta* tolerant, Identified for UP, Bihar, Jharkhand, Bihar, Andhra Pradesh, Orissa, Chattisgarh, Madhya Pradesh and Maharastra by AICRP-VC.

16

Dolichos Bean

Dolichos bean is known as Lablab bean, hyacinth bean or sem. Grown as field crop in MP, Maharashtra, AP and TN. Genus *Dolichus* is renamed as *Lablab* (Syn. *D. lablab, D. purpureus, Lablab niger, L. bvulgaris)* and *Lablab purpureus* name is *widely accepted. Dolichos lablab* L. (synonyms *Lablab purpureus* (L.) Sweet.) is leguminous pod vegetable grown in tropics and few temperate countries. *Lablab niger* var. *lablab is* synonym to *Dolichos lablab* var. *typicus* where long axis of seed is parallel to suture and is vegetable type. *Lablab niger* var. *lignosus* synonym to *Dolichos lablab* var. *lignosus* where long axis of seed is perpendicular to suture and is pulse type.

Verdcourt (1970) classified the species in to 3 sub species

1. *Lablab purpureus* subsp. *purpureus:* Common cultivated races of hycianth bean.
2. *Lablab purpureus* subsp. *Uncinatus :* Slender inflorence with smaller pods (4 cm length, 1.5 cm width).
3. *Lablab purpureus* subsp. *bengalensis*: Pods are linear, similar to kidney bean in appearance

These types are interbreed freely.

Botany: It belongs to family *Leguminosae*, subfamily *Papilionaceae*, tribe *Viciae* and genus *Lablab*, *L. niger* Medic. (syn. *Lablab vulgaris* Savi, *Dolichos lablab* L.). Plant is a perennial herb, often grown as an annual; twining or bushy, colouration of stems, foliage, flowers, pods and seeds variable; pod shape variable, leaves alternate; trifoliate, stipules present, inflorescence erect; long racemes, flowers-white or purple; papilonaceous, stamen 9+1; anthers uniform, ovary sessile, ovules ± 6, style incurved, stigma terminal glabrous, capitate, pods oblong flattened, style-persistent.

Pokle and Deshmukh (1971) studied floral biology of dolichos bean where anthesis occurs 9 to 17 hr and anther dehiscence occurs between 5 and 14 hr. Pollen is fertile on the day of anthesis. The two botanical varieties *lignosus* and *lablab (typicus)* are cross-compatible. The chromosome number is 2n=20, 22, 24.

Origin and Distribution: Presumably tropical Asia and probably India is the center of diversity for dolichos bean and later it was introduced to China, Sudan and Egypt. Crop is believed to have an Asian origin and spread throughout the tropical and temperate regions of Asia, Africa and America. The original ancestral species of the crop is considered as *Dolichos purpureus.*

Genetics

Character	No. of genes	Type of gene action
Pigmentation of vegetative	Monogenic	Purple dominant to white
Pod colour	Monogenic	Green dominant to light green
Pod character	Monogenic (FpFp)	Flat dominant to swollen
Seed colour	Monogenic	Chocolate dominant to brown

Studies indicate that pods/plant, seeds/pod and 100 seed weight are highly heritable and highly correlated with yield.

Objectives

1. Bushy habit
2. High yielder
3. Early yielder
4. Photo insensitive
5. High sugar to polysaccharide ratio
6. Less fiber (string less)
7. Fleshy pod pericarp

There is a need to develop bushy types which are early, high-yielding, disease resistant and non-season bound. Earliness, high sugar to polysaccharide ratio, less fibre, stringlessness and fleshy pod pericarp are the desirable characters in dolichos bean.

Breeding Methods: Dolichos bean is a highly self-pollinated crop and cross-pollination through insects has also been reported. The protruded stigma (exerted) in a few varieties leads to allogamy.

Bulk population breeding: This involves growing of genetically diverse populations of self fertilizing crops in a bulk plot with or without mass selection, followed by single-plant selection.

Annappan *et al. (*1989) made local and exotic collections in field *lablab*. Selections were also made from bulk crop in farmers plots based on high phenotypic expression for yield components like branches/plant, number and size of pods. Four selections were made from the above collections. Considering high yield, reduced duration, edible nature of green and dry beans and drought tolerance, 'DL 2539' was released as CO l Mochai'. 'CO1 Mochai' is early, high yielder and drought tolerant cultivar.

Pure line selection and backcross

Arka Jay (Sel-1): Developed from back cross and pure line selection from cross Hebbal Avare x IIHR 93. It is photo insensitive and tolerant to heat and drought.

Arka Vijay (Sel.-2): Developed from back cross and pure line selection from cross Hebbal Avare x IIHR 93. It is photo insensitive and tolerant to heat and drought.

Pedigree method: Ramasamy *et al.*(1990) resorted artificial cross-pollination between two garden bean varieties CO 9' (bushy type) and white 'Yanaikathu' (Pendal type). The hybrid derivative 'Colt 22' was selected during F6 generation. The selected culture was then evaluated under preliminary yield trial, comparative yield trial, multilocation trial and on farm trial.

The culture Colt 22' named as 'Co 11 Avarai' and was released for cultivation.

CO-9 x Yanaikathu
(Bush type) (Pendal type)
F6 Colt 22

Named as "CO 11 Avarai"

Heterosis breeding: Although heterosis is reported for yield and better pod quality (CO5 x DL3196) and ascorbic acid content (6023A x 6010, 16019' x 6010' and '6014 B x 6015) hybrid seed production is not economically feasible. Suggested methods of F1 seed production are hand-emasculation and pollination, making use of protruded stigma and use of selective differential gametocide like FW 450.

Mutation breeding: Sivasubramanian *et al.* (1989) gamma irradiated seeds (24 kr) of a bushy variety 'Co 6'. CO-6 is the derivative of a cross between DL 244' (garden bean) and 'DL 3196' (field bean). Gamma irradiated progenies were screened for green-pod yield and quality. In the M2 generation, a mutant possessing economic values was isolated and designated as 'Mutant 3'. The mutant being higher yielder than Co 9' as well as possessing better values for sensory tests and biochemical analysis, it was released as 'Co 10 Avarai,'

DL 244 x DL 3196
Garden bean Field bean
CO-6

Seeds were gama irradiated (24 Kr)

In M2 isolated 'Mutant 3' which is high yielder than CO-9 and released as " CO 10 Avarai"

Salient Achievements

Pusa Early Prolific: A selection made at the IARI, New Delhi is a promising vegetable type in dolichos bean and is early and has long thin pods in bunches.

Rewa: It has a protein content of 25.11 per cent.

Deepaliwal: Developed at Punjabrao Krishi Vidyapeeth, Maharashtra. A high yielding, pole type variety matures in 91 days.

Varuna: Local variety.

CO-1: Pure line selection released from TNAU during 1990. Plant height 60 'to 70 cm, stem green, leaves trifoliate, pods flat, green when tender and tan colored at maturity. Seeds bold with black colour, yields 16 q/ha in 140 day.

CO-2: Derived from CO-8 x CO-1, grows 60 cm height, photoinsensitive, erect, bushy, 5 to 6 branches, flowers borne on axillary racemes, pinkish purple at blossom, bluish-purple on fading, pods flat, green and tetra seeded with medium ovate shape.

Hebbal Avare-3: Derived from Hebbal Avare-1 x US67-31 and released from UAS, Bangalore during 1978. Plant height 65 to 75 cm, erect with determinate growth habit, photoinsensitive, flowers white in colour, pods green, 2 to 3 seeded, yields 8 to 10 quintals/ha in 90 to 100 days.

Wal Konkan-1: Derived from Wal 2-K2 x Wal 125-36 and released in Maharashtra during 1984, plant height 40 to 45 cm, bunchy with long tendrils, leaves dark green and smooth, seeds bold with brown coat colour, protein content 14.68 per cent. It is resistant to yellow mosaic virus and yields 9 to 10 q/ha in 110 to 115 days.

125-36: A selection from local collection and released in Gujarat. Plant height 75 to 100 cm, stem and foliage green, pods short (3 to 4 cm), medium broad, 3 to 4 seeds per pod, pods borne in clusters, 80 to 100 pods per plant. Seeds milky white and duration is 110 to 120 days.

Hebbal Avare-4: It is soft poded variety derived from Hebbal Avare x CO-8, gives 60 q/ha in 5 pickings. Compact and dwarf plant type, first picking starts within 60 days.

Pusa Sem 2: It has semi flat dark green and non-stringy pods of 15 to 17 cm long. It is tolerant to anthracnose, viruses, aphids and pod borers. It gives 14 to 17 tonnes of yield per hectare under field conditions and 22 to 28 tonnes per hectare on trellies.

Pusa Sem 3: It gives flat, green, very tender, meaty and stringless pods of 15 to 16 cm long. It is susceptible to frost. It yields 14 to 17 t/ha under field condition and 22 to 28 t/ha on trellies.

Dolichos bean (Pole type)

Swarna Utkrisht

- Developed through pure line selection from CHDB-1
- Pod : Straight flat green fleshy (10-12 cm) having very good cooking quality
- Matured seeds are light brown
- Tolerant to anthracnose, bean mosaic virus and pod borer
- Time of sowing: June-July
- Seeds rate: 6-8 kg/ha
- Maturity : First harvest 110-120 days after sowing
- Fresh pod yield : 35-40 t/ha
- Recommended for Uttar Pradesh, Jharkhand, Bihar, and Punjab

Arka Jay: (Sel. 1): It is developed through back cross and pure line selection involving parents Hebbal Avare X IIHR 93. Plants dwarf and bushy and photo insensitive, flowers purple, pods long, light green slightly curved and without parchment. It is vegetable type with excellent cooking qualities. Tolerant to low moisture stress and gives yield of 12 t/ha in 75 days.

Arka Vijay (Sel. 2): It is developed through back cross and pure line selection involving parents Hebbal Avare X IIHR 93. Plants dwarf and bushy and photo insensitive, flowers white, tolerant to heat and drought

Arka Amogh: Plants are medium tall, for 50% flowering takes 40 days and pods are ready for harvest in 55 days. Pods are similar to Arka Jay and Konkan Bhushan. It yields 19-20 t/ha.

Arka Sambhram: Plants are medium height, 50% flowering observed in 40 days and pods are ready for harvest in 55 days. Pods are flat, light green, medium long (13-15 cm), medium width (1.5 cm). Yield: 19-20 t/ha

Arka Soumya: Plants are medium tall, 50 % flowering in 45 days and pods are ready for harvest in 55 days. Pods are slender (1.0 cm width), medium long (13-15 cm). It yields 19 t/ha.

Kashi Harittima: It is suitable for sowing in kharif season. High yielding and possess good pod quality (green, tender and parchment free). It is moderately resistant to Dolichos Yellow Mosaic Virus diseases under field condition. Moderately tolerant to jassid, aphid and pod borer under field condition.

Other cultivars: JDL-37, JDL-79, JDL-53, T-1, HD-18, HD-60, K-6802

17

Alliums (Bulb Crops)

Onion

Onion is very popular vegetable crop grown worldwide. Consumption is in the form of fresh, frozen, and dehydrated bulbs and green bunching onion. Origin of *Allium cepa is* still somewhat mystery. Vavilov (1951) suggested that the onion originated in the area of Pakistan. Others have suggested Pakistan, Iran & the mountain areas to the north (Jone& Mann, 1963). One fact is certain that the onion has been around in its present edible form for thousands of years. Onion recorded in tombs as early as 3200 B.C. It was mentioned as a food in the bible and in the Koran with reference thought to be around 1500 B.C. In addition to being a food considered by people, plant is having certain healing power. Some early drawing indicates that onions were used in spiritual offerings.

Allium Species

1. ***Allium Porum*:** Leek- non bulb forming.
2. ***Allium cepa*:** Includes bulb type onion, multiplier onion & shallot of Europe & US. Shallots are different from multiplier onions by way of forming into single bulb followed by division & their top dies indicating maturity. They go into state of rest or dormancy similar to common onion. Shallots crosses freely with common onion & it produces fertile progeny. They will be treated as *Allium cepa* L. closely related.

   India: *Allium cepa* var. *cepa* – propagation by seeds and single bulb.

   Allium cepa var. aggregatum. Propagation by bulbs which are in clusters.
3. ***Allium ampeloprasum*:** These are highly variable, bulbing and non bulbing types existing. Great headed garlic, leek, & kurrat are common names. The scape (or seed stem) is round and solid. The umbels produce seed and it rarely produces bulbils and sterility common. A leek produces no bulb. It is propagated by seed. Kurrat is a leek like form of small stature. It is raised for its edible tops and is also propagated by seed.

Rakkyo *(Allium chinense G. Don)*: It is native to China & Eastern Asia and is grown by home gardens mainly of Chinese and Japanese background. It is like Chive but it bulbs and divides like multiplies. Seed stem are solid, leaf blades are hallow. Flower is purple in colour but do not set seed, because they are tetraploid (Kurita, 1952).

Japanese bunching or welsh onion (*Allium fistulosum* l.): Popular in China and Japan. Flower open fully at anthesis and open from central flower to out ward to umbel base. Shape of seed stem in *Allium cepa* is swelling in mid way and in *Allium fistulosum* it is straight and uniform (seed stem).

Garlic: (*Allium sativum* L.): Leaves flat, longitudinally folded leaf blade very similar to *Allium ampeloprasum.* Flowering plants produce inflorescence having small bulbils instead of flower. Some flower may partially develop but produce no seed. The bulb is a composite of several cloves. Garlic propagated by planting individual cloves. Great headed garlic (*Allium ampeloprasum*) has single large bulb and with smaller bulbs or cloves attached where garlic has same sized cloves in a bulb.

Chives (*Allium schoenoprasum*): The chive is probably the most variable in type and range. Its flowers, usually purple in colour, open first at the top and continue to base of the umbel. This and *Allium fistulosum* have same opening pattern. It has slender leaves, thereby forming dense clumps. Only leaves are eaten and not bulbs.

Chinese chives (*Allium tuberosum*): Common in China and Japan for its edible leaves and young flowers. It produces enlarged rhizomes but they are not for food. Propagation is by division of clumps of rhizomes. It has grass like leaves. The scape is solid and has two or more sharp angles running the entire length of the stem. The flowers are white and borne on flat topped umbels and generally there are few flowers per umbel.

Floral biology: (*Allium cepa)*: Temperature influence flowering but not day length. It bears few to more than 2000 flowers per umbel. The flower stalk (scape or seed stem) where internodes extending the stem. Number of scapes produced per plant depends on the number of lateral bud contained on the stem. Plant grown from seeds generally produces only one scape (seed stem). Since several lateral buds may be present that formed during development of bulb. Plant grown from bulb may produce six or more seed stems.

Umbel: It is an aggregate of many small inflorescences (cymes) of 5 to 10 flowers.

Flower: It bears 6-stamens, 3-carpels which are united and 6-perianth segments.

Pistil: Contain 3 locules, each of contain 2 ovules, flower also contain nectarines.

Anthers: Shedding of pollen takes place over period 3or 4 days prior to the time when style attains full length. At this time stigma because receptive i.e. protandry (cross pollination favorable).

Capsule: It is dry and split from the apex.

Disorders in flower: Jones and Emsweller (1936) reported male sterility in variety Italin red. Jones and Clarke (1942) identified genetics of male sterility. Where N/S-cytoplasmic factor and msms as nuclear factor and S msms genotype gives male sterile plant and Nmsms genotype gives male fertile plant and similarly S Ms ms as heterozygous male fertile plant and S Ms Ms as homozygous male fertile plant.

Pollination control: For normal plant selfing can be done up to S2 generation (floral wise inbreeding depression) by bagging.

Crossing

1. Hand emasculation of stamens very difficult and continuous flowering make task more difficult.
2. Put two plants to be crossed into a cage and the house flies, blow flies, or bees and identify F1 at bulbing stage. F1 bulbs identified and put into cage to produce F2 or can be backcrossed. Small cage can be used for making several plants Selfing. Use insects as pollination agents. Nylon screen material can be covered. Big cages may be used up to 100 ft long and 12 ft wide. Otherwise isolation distance of one or more miles can be followed. Male sterility is another factor which can be used in crossing.

Major Constraints for Onion Production and Productivity

1. Low yielding varieties having poor quality bulbs and poor storage life.
2. Non-availability of suitable F1 hybrids has been one of the major constraints in improving the productivity.
3. Diseases like purple blotch, basal rot, stemphylium blight and storage rot besides insect pests like Thrips. Available varieties are susceptible to these diseases and Thrips.
4. Lack of suitable varieties for dehydration.
5. Lack of varieties suited for export market.
6. Non-availability of varieties resistant to abiotic stresses

   Moisture stress, High temperature, Salinity, Alkalinity

Major Objectives

1. Develop high yielding varieties having better quality bulb with longer storage life and suitable for growing under varied agroecological conditions.
2. Development of F1 hybrids with better quality bulbs & resistant to major diseases.
3. Development of varieties resistant to major diseases like purple blotch, basal rot, stemphylium blight, bacterial storage rot and insect pests like Thrips and aphids.
4. Breeding white varieties with high dry matter content suitable for dehydration industry.
5. Development of varieties suitable for export market.
6. Development of varieties resistant/tolerant to abiotic stresses like moisture stress, high temperature, soil salinity and alkalinity.

Leading Centers Working on Crop Improvement in India

1. ICAR, Directerate on Onion and Garlic, Rajguru Nagar, Pune
2. IARI, New Delhi
3. IIHR, Bangalore
4. Mahatma Phule Krishi Vidyapith, Rahuri.
5. Associated Agriculture Development Foundation, Nasik.
6. NBPGR, New Delhi.
7. PAU, Ludhiana.
8. HAU, Hissar.
9. VPKAS, Almora.
10. TNAU, Coimbatore.

AICVIP -18 Centers

Genetic Resources

NBPGR has more than 1000 onion accessions and other Research Institutes/ units also have sizable amount of germplasm. Indigenous collection account for more than 95% of the germplasm lines maintained in onion, there is need for importing more exotic collection.

NBPGR: High TSS up to 20% – 6 Lines.

Compact Large Sizes Bulb – 15 lines.

Moderate resistant to *Stemphyllium* blight – 50 lines.

Resistant to Purple blotch – 30 lines.

Tolerant to both the diseases – 16 lines.

Sources also identified for resistance to purple blotch, basal rot, and thrips, tolerant to soil salinity, tolerant to bolting tendency, good keeping quality besides high dry matter production.

Cytogenetics: All cultivated *Alliums* have same basic chromosome number x =8, *Allium cepa* is a diploid have 8 pairs of metacentric or submetcentric chromosomes. Natural triploid-variety *Pran* of Kashmir belong to *Allium cepa* var. *viviparum*. (Singh et.al, 1967). Variety Pran has mixed heads of both flowers & bulbils (Kaul &Gohil, 1974). *Allium fistulosum* has 2n =32. Interspecific crosses between *cepa* and *fistulosum* has lot of variation and vigorous growth but, no bulbing and especially abnormality like multivalent, univalent, bridges, and fragments besides heteromorphic bivalents. These abnormality led into very high pollen sterility (Pathak, 1988).

Genetic Studies

Male sterility: Cytoplasmic male sterility (S msms) **identified by** Jones and Emsweller (1936). Another form 'T' cytoplasmic factor control male sterility (Schweisguth, 1973). Male sterility reported in 'Nasik White Globe' was very strong and there was no restorer present in 400 germplasm lines including exotic accessions assed (pathak *et al.*, 1987).

Bulb colour: Four colour classes (white, yellow, and red, brown) of bulb are controlled by series of complementary epistatic genes viz., C, I, R, & G genes (El-shafie and Davis, 1967). As per Patel *et. al* (1971), there are four factors which control skin colour viz., two basic factors C1 and C2 and two duplicate factors R1 & R2 for the expression of red colour.

Yellow x yellow
↓
Pink (Jones & Peterson, 1952) unable to explain the situation.

In general as per Clarke, Jones & Little, 1944

1. II colour inhibitor gives white colour to the bulb irrespective of other genes.
2. Ii off-white or buff coloured bulb.
3. ii other colour genes express example C/- and R/ -
4. cc recessive white condition regardless of R- Gene

Pink produce when homozygous yellow lines crossed-it cannot be explained and indication of existence of complementary genes that modify bulb colour. There is possibility of existence of many complementary and additive genes determining onion bulb colour. There are shades of yellow, red and white and almost like quantitative.

However Following Gentics Information Explain Partially

Homozygous red	ii, CC, RR
Heterozygous red	ii, Cc, RR
-do-	ii, CC, Rr
do	ii, Cc, Rr
Homozygous yellow	ii, CC, rr
Heterozygous Yellow	ii, Cc, rr
Homozygous recessive White	ii, cc, RR
do	ii, cc,Rr
do	ii, cc, rr
Homozygous dominant white	I / I,—,—
Heterozygous dominant white	I / I,—, — anything either dominant or recessive C & R genes.

Other Recessive Monogenic Traits

Albino seedling	a /a
Yellow seedling linked with glossy	y1 /y1
Yellow seedling not linked with glossy	y2 /y2
Pale green seedling	pg / pg
Glossy foliage	gl /gl
Exposed anther	ea /ea
Yellow anther	ya / ya
Pink root résistant	pr / pr

Quantitative Traits

Bulbs yield	Additive & non additive
Bulb shape	Non additive
Bulb size	Additive gene.
Time of maturity	Additive Gene
TSS	Non additive
	Both additive & non additive.
Dry matter content	Additive gene

Bulb yield positively correlated with bulb dry matter and bulb weight and leaves per plant and it is negatively correlated with TSS. Seed yield had positive correlation with number of seed stalks, umbel diameter, seed weight and number of leaves. Seed yield had negative correlation with days to flower. High dry matter, high TSS, high ash content and high reducing sugar associated with longer storage life (Patil & Kale, 1985). Bulb with thick neck, more number of splittings and premature bolting associated with poor storage.

Breeding Procedure and Methods

Long day onions require 14 or more hrs of day length for bulbing and short day onions require 11.5 hrs or less day length and intermediate or day neutral type also exist. Temperature also alters bulbing. Western countries onions which are long day type are storage onions and should be stored under similar condition (cool & dry) as of commercial crop cultivation. Short day type are fresh market onions and should be stored under warm and humid conditions. Long or short day types are suited for dehydration. In India short day prevail in winter (Rabi) follows dry hot condition and can be stored for longer period. Long day prevail in rainy season (*Kharif*) follow rain, cool and humid condition which facilitates sprouting leading to poor storage. However, it is possible to develop good storage variety of short day. Methods of breeding followed are mainly mass selections, heterosis breeding and back cross breeding to transfer male sterility. It is highly cross pollinated crop and hence, mass selection, pedigree methods, bulk method and heterosis breeding are feasible.

Mass Selection

Arka Kalyan-mass selection from IIHR-145

Arka Niketan-mass selection from IIHR-153

Arka Pragati-mass selection from IIHR-140

Arka Bindu-mass selection from IIHR-402

Punjab Selection and Hissar-2 are developed through mass selection.

Pedigree method: Arka Pitambar: Pedigree selection from cross U.D. 102 x IIHR-396 and it is tolerant to Purple blotch, basal rots and thrips. It has long storage life and suited for export.

Interspecific hybridization: To generate variability for yield, dry matter, quality, storage properties, minimum bolting and resistance to diseases resorted to interspecific hybridization. Successful crosses are *Allium cepa* var. *cepa* x *Allium fistullosum* (welsh onion, resistant to thrips) and *Allium porrum* x *Allium cepa* var. *cepa*. Other related species which can be attempted are *Allium vavilovii*, *Allium aschaninii*, *Allium Pskemense* & other species.

Heterosis Breeding

Jones & Clark discovered male sterility in Italian Red 13-53 (specific pedigree entry), but in India ms lines development work is poor & use of exotic ms line is difficult because they are not suited to our photoperiodic conditions. In India male sterility is reported by Patil *et al.,* 1973 in cv. Niphad 2-4-1, Pathak *et al.* (1980) in cv. Nasik White Globe, (IIHR 20) and Netrapal *et al.* (1986) cv. Pusa Red. Strong cytoplasmic factor isolated from cv. Bombay White Globe (Pathak *et al.*, 1986). Using this factor identified F1 hybrids yielding 35% more than best parent. Those are ms1 x NER-1, ms1 x IIHR 21-1, ms8 X IIHR 52-1(Pathak *et al.*, 1987). Joshi & Tandon (1976) reported heterosis up to 72% over MP & 37% over BP for bulb yield.

AT IIHR Developed CMS based F1 hybrids

Arka Kirthiman: F1 hybrid involving parents MS-65 and Sel 13-1-1, tolerant to purple blotch, basal rot & Thrips, long storage, red colour.

Arka Lalima: F1 hybrid with parents MS-48 and Sel 14-1-1, red colour, tolerant to purple blotch, basal rot & Thrips, long storage.

Schematic system for breeding improved open pollinated cultivars

Year	Procedure
1	Grown Source Lines & Select 100 Bulb, Store Bulb
2	Plant selected bulb & self
3	Plant S1seed in progeny rows, selected best bulbs from best progeny row discard poor progenies completely. Store bulbs
4	Plant S1 bulbs & self 5-10 selected bulbs from each progeny. -also make a few 3-5 plant masses from same progeny rows.
5	Plant S2 seed & 3 or 5 bulb mass seed obtained in 4 th year. Select line that look best; if any 3to 5 plant masses look good & are uniform selected these line also. - store bulbs
6	Plant S2 bulbs, i.e. mass 12-15 bulbs from selected progeny row.
7	Plant seed to begin observation trial testing & early evaluation of bulbs. - selected superior progeny & discard other.
8	Mass 100 bulbs & plant in 9 x 10 feet cage for small seed .Increases of selected line.
9	Plant seed for yield trials & in small planting on Commercial farmer, Several location testing. Select stock bulbs for seed increase.
10	Plant bulbs to make a 12x 24 feet cage for seeds increase, serve observe seed yield.
11	Plant several commercial planting to evaluate for all requirements such as shipping, storage & processing ability character this was not possible during earlier testing.
12	Release superior line as a new cultivar.

A procedure to develop new-male sterile onion breeding lines (a line) use the test cross & back cross methods

Year	Known A- line bulb	Fertile selection bulb
1	Plant bulb	Plant bulb, self & cross to A-line
2	Plant F_1 seed	Plant S_1 seed
3	Plant F_1 bulbs (stored at 40°f), -if all sterile make first back cross, -if not 100% sterile –discard.	Plant S_1 bulbs along side F_1 bulbs Massing to avoid inbreeding depression.
4	Plant BC_1 seed to produce BC_1 bulbs.	Plant seeds from small mass of B-line selection to produce S2 B-line bulbs, make selection for type desired.
5	Store bulbs at 40°F and plant and observe flower for male sterility, all should be sterile. Cage with B-line.	Procedure same as A-line. Continue up to fifth back cross
	Continue up to fifth back cross new A- line is synthesized	B line, maintainer for A line is ready.

A Schematic system for breeding improved onion hybrids

Year	Procedure
1	Grow & select 100 bulbs, store, & plant, also grow/get (supply of) male sterile bulbs for use in test crosses.
2	Self selected bulbs & at the same time, test cross with known sterile line.
3	Grow out bulbs from self & F_1 test crosses, select, discard poor progeny rows & their F_1 pair, store only selected progenies bulbs.
4	Plant bulbs for seed production; observe sterility character in F_1 lines, if 100% sterile, self selection & make a back cross to F_1, discard pairs with fertile F_1 lines.
5	Grown out A and B lines as pairs, continue to select in B line side, save best bulbs from A line for next back cross.
6	Self B line of selected progenies & make back cross to sterile lines.
7	Grow bulbs & make final selection on basis of B line side.
8	Mass B line using 10-20 bulbs in cage while making the second back cross to the sterile side of the pair.
9-12-Sep	At this point, begin going seed to seed and continue through fifth back cross using the same procedure as in 8th year; many A& B lines should be developed by this time; begin making hybrid combinations for testing.

Disease Resistance Breeding

Purple blotch: ***Alternaria porri***: More series in *kharif* (62% incidence) then in Rabi (38% incidence) (Quadri *et al.*, 1982). As small sunken whitish flecks with purple colour centre later enlarge & form large purple area.

Resistant: IHR 56-1 (Pathak *et al.*, 1986)

Moderately resistant: Red Creole (Sandhu *et al.*, 1981), Pusa Red (Raju & metha, 1982) and Pusa Ratnar and Arka kalyan (Pathak *et al.*, 1986)

Tolerant: Rampur local, Pusa Red, Patna Red (Raju & Raj, 1979) and IC32176 (Bisht *et al.*, 1990)

Stamphyllum blight: It is caused by *Stemphyllium vesicarium Wallrotn* where 40 accessions found resistant at NBPGR (Bisht *et al.*, 1990).

Tolerant: IC32176 (Bisht *et al.*, 1990)

Basal rot: (*Fusarium oxysporum* f sp *cepae* Schlect)

Resistant: Telagi Red, Poona Red, Bellary Red, Patna Red and White Large (Sokhi *et al., 1974)*

Basal rot: more common in Bihar near Nalanda district (Gupta *et al.*, 1983), genotypes showing resistance at seedling stage were found susceptible later at bulb development stage (Sokhi *et al.*, 1974).

Other diseases are: Downey mildew caused by *Pernospra* sp is found on Hill tracts & plain on seed & bulbs crop.

Insect resistance: Thrips are major pests.

Thrips: All over India it is spotted. Symptoms first appear on leaves which subsequently turn into pale white blotches due to the sucking of sap.

Sources: Varieties identified as resistant: N-2-4-1, Sel 104, Pusa Ratnar, Kalyanapur Red Round (Singh *et al.,* 1986)

Pawar *et al.,* 1986: White Creole, Kagar-2 & Peta-1

Tolerant: Hissar-2, Panipat Local, Rajapur Red, N-53, Bombay white, Pusa Ratnar, (Kaur *et al.*, 1979)

Lowest incidence: *Allium fistullosum* BS 956 (Mote & Sonane 1997)

Tolerant varieties: Hissar-2, Bombay white (Saxena, 1977)

Head Borer: *Heliothis armigera*

Seed crop heavy infestation (Raj *et al.*, 1979) bulb crop feeding on leaves (Bhardwaj *et al.*, 1983)

Least preference : Sel 102-1, Large Red, Verna Giant (seed crop) (Singh *et al.,* 1983).

Breeding for Processing Qualities: Dehydrated products such as flakes, rings, granules, powder etc. and processed onion as onions in vinegar and brine are the important byproducts being prepared and marketed worldwide. Processing industries in any commodity play an important role in stablizing prices in domestic markets. Dehydration industries demand for white onion varieties with globular shape bulb and high TSS (> 18%). All Indian white onion genotypes are having TSS range between 11-13%. Kataria (1990) reported three white mutants developed through chemical and physical mutagens viz., 22-5-1-1, 22- 9-2-2 and 106-13-1-1 having TSS range from 25-30 per cent. This excellent material never reflected further in the form of either commercial variety or breeding material for varietal improvement. Present day industry is pulling on with the help of meager varieties and local genotypes available for processing. V-12 variety of white onion recommended and used by Farm-Fresh-Jam Irrigation Systems Limited is claimed to have 18°B TSS. Development of white onion varieties with high TSS has yet remained a classical agenda on onion breeding programme.

Breeding for Export Quality Onion: India is number one in export of onion followed by Netherlands. India's export is mostly to South East Asian and Gulf countries. Dark red and light red onions with globose shape are mostly preferred with various diameter sizes. The present practice of export is grading and packing from the total bulk arriving in various onion markets. Uniformity is shape size and colour is seldom attained, as there no systematic control over planting of required varieties. Further, there is lack in varieties, which can suit to exclusive markets. European markets require yellow or brown onion with big size. There are hardly any indigenous varieties, which can meet to these standards. NRCOG has initiated work in this direction and recommended Mercedes from exotic material for growing in late kharif season.

Varieties Developed

Pusa Red: It is selection from indigenous collection, released in 1975. Bulb is 70 to 90g in weight, 5 to 6 cm diameter, less pungent and can yield 20 to 30t/ ha in 125 to 140 days after maturity and it can produce 6 to 9 leaves /plant.

Arka Niketan: It is developed thorough mass selection from local collection (IIHR153) released in the year 1987. It produces 8 to 10 leaves / plant, globbose with thin neck, bulb weight is 100 to 180 g, 12 to 14 % TSS and it is highly pungent. It has good storage life (5 months), yields 33 t/ha, duration is 140days and suited for Rabi & kharif seasons.

Arka Kalyan : It is developed through mass selection from local collection (IIHR145)-kalawana tai mahour. It has globular, deep red bulbs with 100 to 190 g bulb weight and can yield 34 t/ha and duration is 140days. It is moderately resistant to purple blotch and suitable for kharif season.

N-2-4-1: Released from Mahatma Phule krishi vidhyalaya, Rahuri in Maharastra. Bulbs light red, globe shaped with 6.1 cm diameter and 11 % TSS and is pungent. It gives 25 to 30 t/ha in 140days and has good keeping quality.

Punjab Selection: Developed from PAU through mass selection from indigenous material, released in the year 1975. It produces 6 to 9 leaves and bulbs are globular, red in colour and 5 to 6 cm in diameter with 50 to 70g bulb weight and have good keeping quality. Its yield potentiality is 30 t/ha.

N-2-4-1: This variety was developed at MPKV, Rahuri. Plants are of 40-45 cm height. Bulbs are globular (6.1 cm diameter), brick red; 12-13% T.S.S., pungent and firm with good keeping quality; suitable for rabi season; gives an average bulb yield of 30 t/ha in 140 days of crop duration; Recommended for release and cultivation in 1985 for the zone IV (Punjab, U.P. and Bihar) and VII (M..P. and Maharashtra).

Arka Niketan (Sel-13): Arka Niketan was developed at IIHR, Bangalore, derived through mass selection from a local collection (IIHR-153) from Duberd of Maharashtra. Plants are of 6-70 cm height with 8-10 leaves per plant. Bulbs are globular, pink outer scales arranged in tight concentric ring, weight of 100-180 g, highly pungent, 12-14% T.S.S. thin neck; good storage quality, can be stored up to 5 months at ambient temperature; suitable for both rabi and kharif season cultivation; gives an average bulb yield of 42 t/ha in 145 days of crop duration and seed yield of 0.8 t/ha. Recommended for releade and cultivation in 1987 for the zone VII (M..P. and Maharashtra).

Agrifound Dark Red: This variety has been developed at NHRDF, Nasik; derived through mass selection form a local stock of kharif onion. Bulbs are dark red, globular, 4-6 cm in size, moderately pungent with 12-13% of T.S.S. and thin skin; average keeping quality; gives 30-40 t/ha bulb yield in 95-100 days after transplanting. Recommended in 1987 for release and cultivation during kharif season for the zone IV (Punajab, U.P. and Bihar).

Arka Kalyan (Sel-14): Arka Kalyan has been developed at IIHR, Bangalore; derived through mass selection from a local collection (IIHR-145) from Kalawana Taluk of Maharashtra. Plants are of 70-90 cm height. Bulbs are globular, pink flesh succulent concentric internal scale, weight of 130-190 g, 11-13% T.S.S.; moderately resistant to purple blotch; suitable only for kharif season cultivation; gives an average bulb yield of 47 t/ha in 140 days of crop duration and seed yield of 0.8 t/ha. Recommended in 1987 for release and cultivation for the zones IV (Punjab, U.P. and Bihar), VI (Rajasthan, Gujarat, Haryana and Delhi), VII (M.P. and Maharashtra) and VIII (Karanataka, Tamil Nadu and Kerala).

Agrifound Light Red: This variety has been developed at NHRDF, Nasik; derived through mass selection form a local stock of rabi season cultivar, grown in Dindori area of Nasik. Bulbs are light red, globular, tight skin, 4-6 cm in size with 13% of T.S.S.; good keeping quality; gives bulb yield of 30-32.5 t/ha 160-

165 days after sowing. Recommended in 1995 for release and cultivation during rabi season for the zone VII (M.P. and Maharashtra).

VL-3: This variety was developed at VPKAS, Almora (U.P.). Plants are 50-70 cm tall with deep green waxy foliage. Bulbs are red, medium, globular and pungent; gives and average bulb yield of 25 t/ha in 145 days of crop duration. Recommended for release and cultivation in 1990 for Uttaranchal.

Punjab Red Round: Punjab Red Round is an early maturing varietly, developed at PAU, Ludhiana. Bulbs are shining red, globular, medium with thin neck; gives bulb yield of 28-30 t/ha. Recommended in 1993 for release and cultivation during rabi season for the zone IV (Punjab, U.P. and Bihar).

Punjab Naroya (PBR-5): PBR-5 has been developed at PAU, Ludhiana; derived through selection from a line collected from Maharashtra. Plants are medium-tall and green. Bulbs are red, medium to large, round with thin neck; gives an average bulb yield of 37.5 t/ha in 123 days after transplanting; tolerant to purple blotch, thrips and heliothis. Recommended in 1997 for release and cultivation for the zone VI (Rajasthan, Gujarat, Haryana and Delhi).

Pusa Ratnar: Pusa Ratnar was developed through single plant selection at NBPGR, New Delhi. Plants are 30 cm tall and dark green leaves with waxy bloom. Bulbs are bronze deep red, more exposed (above the ground at maturity), obviate to flat-globular, less pungent and drooping neck with good storage quality; gives bulb yield of 32.5-35.0 t/ha in 145-150 days of crop duration. Recommended for release and cultivation in 1975 for plains Indian.

Pusa Ratnar: It is developed at NBPGR and released during 1975. Leaves are dark green, bulbs more expend and bronze deep red, flat globular, less pungent and has drooping neck with good storage quality. It yields 32.5 to 35 t/ha. in 148 -150days duration.

Agri Found Dark Red: It is released from Associated Agricultural Development Foundation, Nasik (now NHRDF) during 1987. Developed from local stock and suited for kharif. Bulbs deep red globular, moderate pungency and has 12 to 13% TSS. It can give 30 to 40 t/ha. in 130 to 140days.

Arka Pragati: It is developed from IIHR 149 stock from Nasik and released during 984. It has bulbs with globe shape, thin neck, highly pungent, early and suited for Rabi and kharif seasons. It can give 33 t/ha in 120days.

Hisar-2: Released from HAU developed through mass selection by using indigenous material. It produces bronze red, flatting and globular shape bulbs. It can give 30 t/ha yield in 130days. It is tolerant to salinity.

Kalyanpur Red Round: It is popular in UP, developed from C.S Azad University of Agriculture & Technology, Kalyanpur. It produces bronze red, globular shape bulbs with moderate pungency and has 13 to 14% TSS. It can yield 25 to 30 t/ha in 130-150 days.

S-148: This variety was developed at PAU, Ludhiana. Bulbs are white, flatish-round, average weight of 80 g, T.S.S. 12-13%; good storage quality; gives bulb yield of 25-30 t/ha in 140 days of crop duration. Recommended in 1975 for release and cultivation in rabi season for the zone IV (Punjab, U.P. and Bihar).

N-257-9-1: This variety has been developed at MPKV, Rahuri; derived through selection from a local collection. Bulbs are globular and white; suitable for rabi season cultivation; gives an average bulb yield of 25 t/ha in 120 days of crop duration. Recommended for release and cultivation in 1985 for the zone VII (M.P. and Maharashtra).

Arka Swadista: It is a white onion variety for fermented preservation and has uniform white color bulbs which are oval to globe in shape and has TSS of 18 to 20 %, dry matter content of 15 to 18%. Its edible bulb is 98% of total bulb. Bulb weight is 35 to 40 g which is small in size with 3 to 3.5 cm in diameter. It is suitable for bottle preservation. It gives bulb yield of 16 to 18 t/ha in 105 days.

Garlic Improvement: Garlic is one of the important edible Allium crops grown and used as a spice or a condiment throughout India. It is also important foreign exchange earner for India. It is rich in proteins, phosphorus, potash, calcium, magnesium and carbohydrates. The garlic bulb contains a colourless odourless water-soluble amino acid allin, On crushing the garlic, the enzyme alliane breaks down aliin to produce allicin of which the principle ingredient is the odoroferous diallyle disulfide. Garlic contains about 0.1% volatile oil, The chief constituents of the oil are diallyle disulphide (60%) diallyle trisulfide (20%) and allyle propyle disulfide (6%). The diallyle disulfide is said to possess the true garlic odour. Allicin present in aquous extract reduces cholesterol concentration in the human blood. The inhalation of garlic oil or garlic juice has generally been recommended by doctors against pulmonary tuberculosis, rheumatism sterility, impotency, cough and red eyes diseases.

History & Origin: Garlic has been cultivated for thousands of years. It is among the most ancient cultivated vegetables. Original abode of garlic is said to be Central Asia and Southern Europe, specially Mediterranion region. It is known in Egypt in Predynastic times before 5000 BC and also to ancient Greeks and Romans. It has long been grown in India and China. It was used in England as early as first half of the 16t century. Most recent researchers consider Central Asia to be the original home of garlic. Some researchers discovered a number of fertile clones of a primitive garlic type, on the northwestern side of the Tien

Shan Mountains in Central Asia. Etoh concluded that this area was the centre of origin of garlic. Gvaladze (1961) in Georgia, proposed a sub-classification of *A. sativurn* into following three groups.

Flowering plants with the bulbils in the inflorescence.

Flowering plants with both flowers and bulbils in the inflorescence,

Plants that form no flower stalks.

Erenburg (cited by Kazakova, 1978) reports flowering and seed production of garlic in Kazakhstan (Central Asia) and Dugestan, port of the Caucasus.

Sources of Genetic Variation: Garlic was probably highly variable in the primary centre of evolution, even before its dispersal from that region. Thereafter intraspecific variation must have increased and isolation must have accelerated diversification presuming that sexual reproduction occurred outside the centre of origin. Today garlic has great variation for maturity date, bulb size, shape and colour, flower and pungency, clove number and size, number of whorl of cloves bolting capacity, scape height, number of size of top sets (bulbils) and number of flowers and fertility (McCollum, 1976). From isozyme and RAPD analysis Pooler and Siman (1993a) and MaaB & Klass (1995) were able to show that great heterogenecity exists within the Central Asian cultivar group. Another RPAD analysis with 72 accessions collected from around the world also showed considerable genetic diversity in the Central Asian Group (Hong 1999). The early domestication of garlic took quite different turn from that of seed propagated leek and onion, Garlic is exclusively vegetatively propagated by cloves or bulbils. Some cultivars are reported to produce flowers but there is no seed seeting. Garlic cultivars differ in maturity, bulb size and number of cloves skin, scale colour, bolting and flowering habits as also pungency and characteristic flavour and aroma levels.

Current Status of Improvement in India: Not much work was done earlier on garlic improvement until National Horticultural Research and Development Foundation started work on development of varieties about two decades ago. The work was also initiated by different state Agricultural Universities and ICAR, Institutes subsequently. The work was further strengthened after the establishment of NRC for onion and garlic by ICAR. Major work on garlic varieties development is being taken up by NHRDF, Nashik, NRCOG, Raigurunagar, HAU Hissar, MPKV Rahuri, PAU Ludhiana, VPKAS Almora, IARI New Delhi, JAU, Junagarh etc. The varieties by different institution and universities are given as under:

S.No.	Name of varieties	Institutes / Universities
1.	Agrifound White (G41)	NHRDF
2.	Yamuna Safed (G-1)	NHRDF
3.	Yamuna Safed -2 (G-50)	NHRDF
4.	Yamuna Safed -3 (G-282)	NHRDF
5.	Agrifound Parvati (G-313)	NHRDF
6.	G-323	NHRDF
7.	Godawari	MPKV, Rahuri
8.	Sweta	MPKV, Rahuri
9.	Pusa Selection-10	IARI, New Delhi
10.	DG-1	RAU, Duragapur
11.	PGS-14	GBPU of agri & Tech, Pantnagar
12.	DARL-52	DARL, Pithoragarh

18

Cole Crops- Cabbage

Classification of *Brassicaceae* family in to different cole crops

Cole Crop: Key to Genera

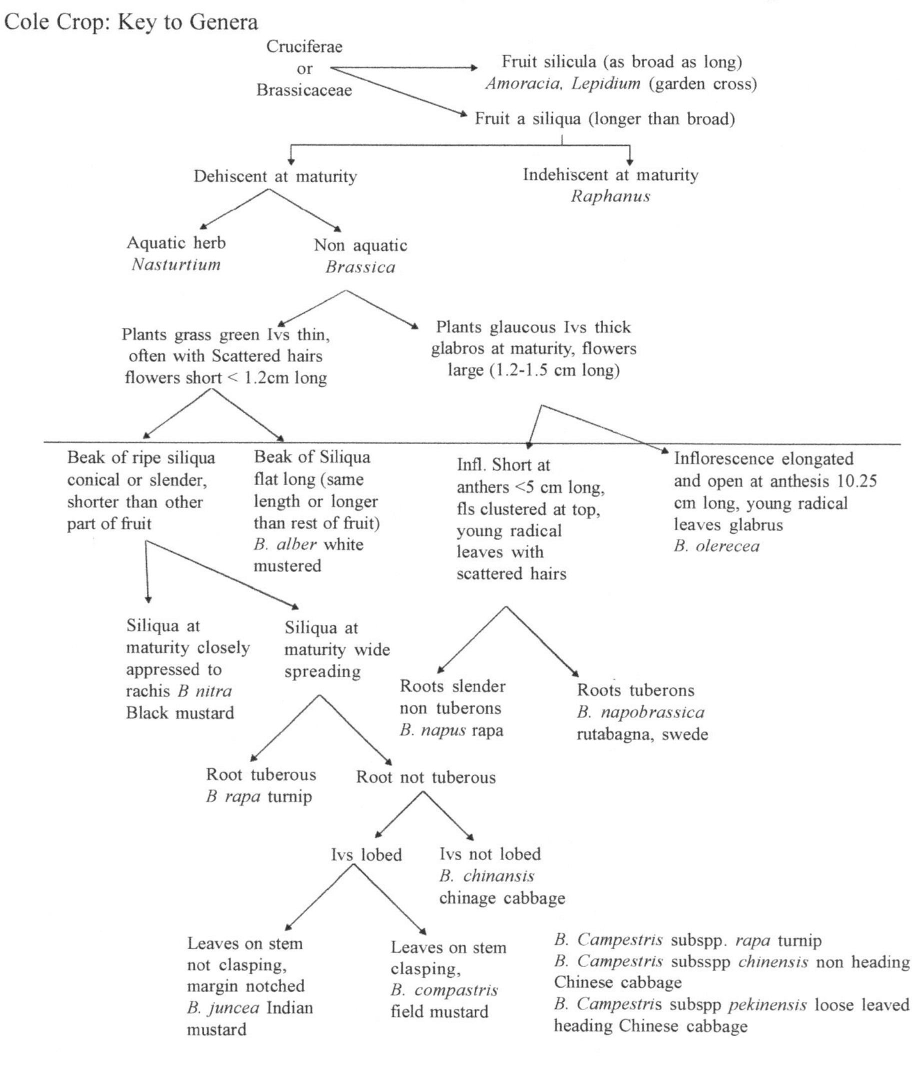

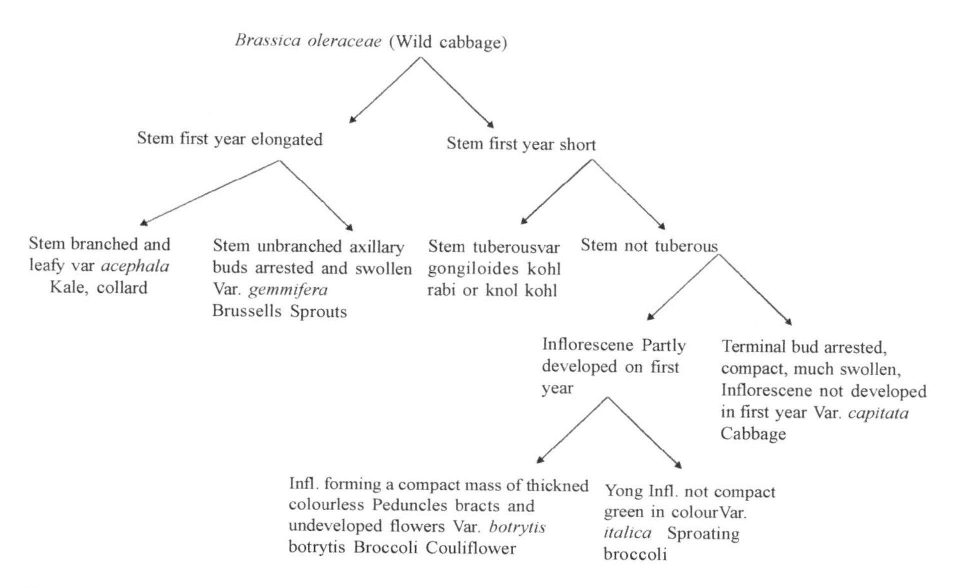

'Cole' abbrivation of word Caulis means stem group of highly differentiated plants originated from single wild *B. oleraceae* Var *B. oleraceae* or *B. oleraceae* var sylvestris Wild Cabbage
Spelt as Kale (English), Kohl (german), Kool (Dutch), Col (Spanish) etc.

Cabbage

Cabbage is an important crop in USA, Northern European countries, Russia, Japan & Australia and its breeding occurs here, but Holland and Japan are the major breeding countries. It is also being cultivated in India and crop improvement work is taken up in India also.

Origin and General Botany: Historical evidence indicates that modern hard head cabbage cultivars descended from wild non headed *Brassica* originated in the Eastern Mediterranean and in Asia Minor. The Celts and Late Romans are responsible for spread cabbage throughout Europe. Latin name *Brassica* derived from Celtic word "*bresic*" meaning cabbage. Over a period hard headed evolved in Northern Europe and loose headed and heat resistant types evolved in South Europe. Cabbage was first introduced to America in 1541 by French explorer Jacques Cartier (planted seeds in Canada) and earlier colonist introduced the crop in to India also. Round headed types-are evolved earlier than flat (evolved in 17th or 18th century). Broccoli, Cauliflower, Brussels Sprout, Collards, Kale, and Kohlrabi are all readily intercrossed members of this species.

Evolution of Amphidiploid *Brassica* sp.
U's triangle proposed by NU in 1935

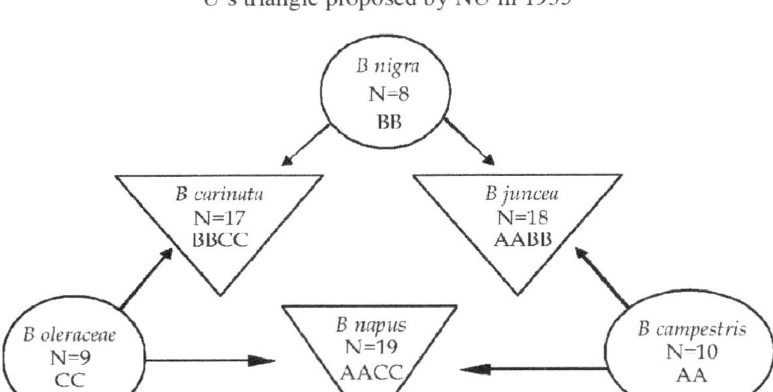

Taxonomy & Cytoplasm Terminology

B. Campestris subspp. *rapa* = turnip

B. Campestris subspp *chinensis* = non heading Chinese cabbage

B. Campestris subspp *pekinensis* = loose leaved heading Chinese cabbage

Usual objective of inter specific crosses is to transfer disease resistance or cms through embryo rescue method. In most of the cases *B. oleracea* has been used as pollen parent successful.

Floral Biology and Controlled Pollination: There are 4 Sepals, 4- petals, 6 stamens and 2 carpel's and ovary is superior. Carpels form a superior ovary with a false septum and two rows of campylotropus (curved) ovules. When the buds are about 5mm long, the megaspore in each ovule divides twice, producing 4 cells, one of which becomes the embryo sac, while the other 3 cells abort. The nucellar tissue is largely displaced by remains of embryo sac. When buds open the ovules mainly consists of two integuments and ripe embryo sac. Androecium is tetradynamous i.e. 2-short, 4- long stamens. Pollen grains 30-40 μm size and have three germination pores. Buds open under pressure of the rapidly growing petals. Opening starts in the afternoon and usually the flowers become fully expanded during the fallowing morning. Bright yellow petals become 15 to 20 mm long and about 10 mm wide. Sepals erect in contrast to the other *Brassica spp*. The anther opens one hour later, the flower being slightly protogynous. Flowers pollinated by insects, bees which collect pollen and nectar. Nectar is secreted by two nectaries situated between basis of the short stamen and the ovary. Base of each of two pairs of long stamen one additional nectary situated, but these two nectaries are not active. Flowers are borne in racemes on the main stem and its axillary branches. The inflorescence may attend the length of 1-2 m. The slender pedicels are 1.5 to 2cm long.

Seed Development: After fertilization endosperm develops rapidly while embryo growth does not start for some days. Still it will be smaller even after two weeks of pollination. It fills most of the seed coat after 3 to 5 weeks. Nutrient reserves for germination are stored in the cotyledons, which are folded together with the embryo, radicle lying between them.

Fruit: The fruits of cole crops are glabrous siliqua 4 to 5 mm wide and sometimes 10 cm long, with 2 rows of seeds lying along the edges of septum (false septum an outgrowth of placenta). A siliqua contains 10 to 30 seeds. The siliqua reaches its full length and diameter 3 to 4 weeks after opening of flower. When it is ripe, the two ovules dehisce. Separation begins at the attached base and works towards the unattached end, leaving the seeds attached to the placenta. Physical force ultimately separates the seeds, usually by pushing of the dehisced siliqua against other plant parts by the wind or by threshing operations.

Male Sterility: Although several recessive monogenic mutants for male sterility reported, it was not significant until Pearson (1972) crossed *B. nigra* with broccoli that cytoplasmic male sterility designated N ps/ps, was obtained in *B. oleracea* and developed in cabbage. Unfortunately, this system was complicated by Petaloidy (Perianth: sepaloid and petaloid) and lack of development of nectaries, male sterile flower that was unattractive to bees. Bannerot *et al.* (1974) crossed cms Radish (R. cytoplasm) from Ogura (1968) with cabbage. In BC_4 he obtained normal plants with 2n=18 that (both crops 2n=18 only) were totally male sterile, the flower having empty pollen grains or vestigial (functionless) anthers. Male sterile plants have problems here like pale or white cotyledon and pale- yellow leaves during plant development and it is prominent at low temperature. These problem are cytoplasmically inherited, various attempts to overcome them are made via cell fusion. All *Brassicas* apparently have no fertility restoring genes for the R cytoplasm type of sterility and act as maintainers. Restorer genes are present in radish. Petals of male sterile flowers are smaller than male fertile flower. Nectaries are usually normal for male sterile plants. R cytoplasm of the male sterile radish has also been combined with *B. campastris* and *B. napus* to obtain cms in these species.

Inbred line	X	R. cytoplasm
B. oleracea		*R. sativum*
(N ms ms)		R ms ms

In Back crosses recurrent parent is inbred line of *B.oleracea*

R ms ms : A- line

N ms ms is a maintainer line or B-line.

Self Incompatibility: Genetically, it is sporophytic system where pollen reaction is determined by the genome of the somatic tissue (of the sporophyte) on which the pollen grain develops. The system of self incompatibility is characterized by the following features.

1. Incompatibility is controlled by one S locus having multiple alleles (50-70 alleles in *B. oleracea*).
2. The reaction of pollen is determined by the genotype of the sporophyte on which pollen is produced and therefore is controlled by two S alleles.
3. All the pollen of a plant have similar incompatibility reaction.
4. The two S alleles may show co-dominance (independent action) or may interact by one being dominant over the other.
5. The independence/dominance relationships of S alleles in pollen and in the pistil may differ.
6. It is usually associated with trinucleate pollen and inhibition of pollen occurs at the stigmatic surface.
7. In contrast of this in gametophutic system of self-incompatibility, the pollen reaction is determined by the genotype of the gametophyte i.e. pollen/egg cell itself and in this system the pollen is binucleate and inhibition of pollen tube occurs in the style.
8. Various kinds of S allele interactions in heterozygous genotype (S_1S_2) with sporophytic self-incompatibility could be as follows :

   Dominance : S1 > S2
   Codominance : S1 = S2
   Mutual weakening : No action by either allele
   Intermediate gradation : 0-100% activity by each allele

Table: Sporophytic self-incompatibility system in a cross S1S3 X S1S2 with different inter allelic interaction.

Pollen reaction	Pistil reaction	Compatibility
Independent	Independent	Incompatible
$S_1 > S_2$	Independent	Incompatible
$S_2 > S_1$	Independent	Compatible
Independent	$S_1 > S_3$	Incompatible
Independent	$S_3 > S_1$	Compatible
$S_1 > S_2$	$S_1 > S_3$	Incompatible
$S_1 > S_2$	$S_3 > S_1$	Compatible
$S_2 > S_1$	$S_1 > S_3$	Compatible
$S_2 > S_1$	$S_3 > S_1$	Compatible

Note: $S_1 > S_2$ Means S_1 is dominant Over S_2

Breakdown of Self-Incompatibility: Various techniques as listed below are available for obtaining a temporary breakdown of the self-incompatibility. For details De Nettancourt (1972) and Chiang *et al.* (1993) may be referred:

1. Bud pollination
2. Delayed self pollination
3. Grafting
4. Heat-shocks
5. Application of carbon dioxide
6. Hormones and protein inhibitors
7. Chronic irradiation
8. Acute irradiation of styles
9. Acute irradiation of pollen mother cells
10. Tetraploidization-haploidization
11. Haploidization-diploidization

Breeding and Utilization of Self-Incompatible Lines: It is easy to breed self-incompatible lines of cabbage through continuous self-pollination and selection. When two self-incompatible lines are used as parents to produce hybrids, the reciprocal crossed seeds can be harvested as hybrids. IN 1950, the first cabbage hybrid in the world was developed in Japan using self-incompatible lines. This hybrid was known as Nagaoka No.1.

The superior self-incompatible lines for hybrid seed production should have following characters:

1. Stable self-incompatibility
2. High seed set after self-pollination at bud stage
3. Favourable economic characters
4. Desirable combining ability

Almost all cabbage hybrid seeds are produced using self-incompatible lines all over the world.

Bud Pollination: The basic seed of parental inbred lines is obtained by hand selfing at bud stage. The bud top is removed by tweezers and strippers to expose stigma which is then pollinated with pollen collected from the same plant/line. The seed production plot of parental inbred lines is covered with net to avoid contamination by bees or other insects. Pollen grains are collected afresh from the opened flowers on the same day. The mixed pollen collected from the same line should be used for pollination to avoid viability depression from continuous selfing. If the bagging isolation is applied to multiply the basic seeds, the flowering branch is covered with paraffin bag/muslin cloth bag before the bud opens. The bud size should not be too small or big. Bud pollination done 2-4 days prior to flowering gives the highest seed set.

Special Considerations for F1 Hybrid Production Plots: Isolation distance of at least 2000 m from cauliflower, kohlrabi, broccoli, kale, brussels sprouts genotypes should be followed. Approximately 15 honey bee boxes/ha plot should be maintained. Building-up framework through appropriate staking to prevent lodging should be done. Synchronized flowering of the parental inbreds should be ensured. Planting ratio of 1:1 for the parental inbreds should be maintained. Although harvested seeds from both parents can be mixed-up, it is better to harvest seeds from both the inbreds separately to improve seed uniformity. When the male and female S- allele specificities are identical, self incompatibility acts to prevent the pollen from germinating on stigma or growing in to the style. By this mechanism, self incompatibility prevents self fertilization. Self incompatibility phenomenon also prevents fertilization in crosses between plants of identical genotypes, and in crosses between plants of near identical genotype.

Assaying Self Incompatibility

Quantifying self incompatibility can be done based on two methods.

1. Based on the number of seeds that develop to maturity after each specific self or cross pollination. Disadvantages of this method are 1) it requires minimum of 60 days time (pollination to seed maturity), 2) Expressing of compatibity or of weak incompatibility, follows reduction in seeds number which is also due to stress (disease, water stress and high temperature or other stresses). Thus seed counts at maturity often do not strictly reflect the intensity of expressed compatibility or incompatibility.
2. Florescent microscopic studies: Pollinate flowers on day one and collect flowers after 16 to 30 hrs of pollination and on the same day excised ovaries are then softened in 60% NaOH and placed in Aniline blue for staining. At about 45 hrs after pollination the stigma and style all squashed on a microscopic slide. The Aniline blue stain accumulates in the pollen tubes and florescence when irradiated with UV light. With appropriate light filter, under fluorescent microscope, the tubes are visible and the background of the stylar tissue is largely unseen.

   In style none or few tubes – Incompatibility
   Many tubes – Compatibility
   Intermediate number of tubes - intermediate strength.

Advantages of this method are within two days one can draw the conclusion and hastens other options being tried.

Identifying S-allele genotypes: Once inbred lines are selected for economic part and resistance to stresses, normally it will be heterozygous for SI and need to be selfed by bud pollination to maintain and seed increase. Simultaneously opened flowers also should be selfed. Measure the intensity of self incompatibility. If there is compatible or weakly compatible, all resultant seeds of this plant may be discarded. Avoid out crossing while selfing. This identification of S- allele will be easy if only two alleles in population. If there are three or more s alleles in a population, then the task is nearly impossible.

Plant – Sa Sb

If (first inbreeding generation) contain Sa Sa, Sa Sb, Sb Sb in the ratio of 1:2:1 then there will be no other genotypes if out crossing is not allowed. Allele 'a' and 'b' are tentative assignments. A group of 11 plants from I_1 population provides 95% probability of having at least one plant of each of the three genotype (SaSa, SaSb, SbSb).

Type of interaction	S-allele interaction		Genotype and phenotype			
	Stigma	Pollen	Female	Male		
I	Dom	Dom	SaSa	SaSa	Sa<Sb	Sbb
				I	C	C
			Sa< Sb	C	I	I
			SbSb	C	I	I
II	Codom	Dom	SaSa	I	C	C
			Sa=Sb	I	I	I
			SbSb	C	I	I
III	Dom	Codom	SaSa	I	I	C
		Sa<Sb	C	I	I	
			SbSb	C	I	I
IV	Codom	Codom	SaSa	I	I	C
			Sa=Sb	I	I	I
			SbSb	C	I	I

Dom : Dominance Co-dom : Co-dominance C - Compatible I - Incompatible

Permanent S Allele Identities: The National Vegetable Research Station (NVRS) at Wellesbourne, Warwick, UK has a collection of all known S alleles, which constitutes the internationally accepted nomenclature. The individual breeder may develop homozygous inbreeds throug.. selfing in bud stage and tentatively allot the genotypes as SaSa or SbSb. If SaSa is demonstrated to be reciprocally incompatible with S3S3 maintained at NVRS, SaSa shall be of S3S3 genotype under international nomenclature. However, breeders can also assign S alleles on their own and maintain the inbreds under the designated S allele nomenclature.

The gene list as summarized by Dickson and Wallace (1986)

Symbol	Character
C	Anthocyaninless
A	Basic anthocyanin colour factor
A^{rc}	Coloured lamina
Gl	Glossy foliage, dominant types produce wax on stems
Sm	Smooth leaves
Pet	Petiolate
K	Dominant factor for heading
W	Wide vs narrow leaf
Dw	Dwarf, short internodes, round leaves
T	Tall
wh	White petal
Cp	Crinkly petal
S	Self incompatibility multi-allelic

Simply inherited character and their gene symbols.

Symbol	Character description
K	Dominant factor for heading
W	Wide leaf
n-1, n-2	Recessive factor for heading Substitutes: K-I, K-2, K-3: Loose head of Savoy cabbage in recessive to the head of smooth leaf.
wh	White petal

Genic Male Sterility

ap	Aborted pollen
ms-1	Broccoli
ms-2	Brussels sprout
ms-5	Cauliflower

Cytoplasmic Male Sterility

CMS ms	cyt. R-rad
ms	cyt. N-nigra

Resistance to black rot: Major gene 'f' plus 2 modifiers (1. dominance 1. recessive)

Branching: I and L: inhibit lateral branching and encourage central inflorescence

I: alone enhances lateral branching

Classification of Variability

Based on Colour and form of heads

1. White cabbage : *B. oleracean* var. *Capitata* L. *f. alba*
2. Red cabbage : *B. oleracean* var. *capitata* L. *f. rubra*
3. Savoy cabbage : *B. oleraceae* var. *capitata* L. *f. Sabauda*

Based on Place of Origin

1. Mediterranean cabbage : *B.o.* var. *capitata* ssp. *mediterranean*
2. Oriental cabbage: *B.o.* var. *cap.* ssp. *orientalis*
3. European cabbage. *B.o.* var. *cap.* ssp. *europea*

Based on Head Shape

1. Round shape: Eg. Golden Acre, Pride of India, Copenhagen Market
2. Flat head / drum head: Eg. Pusa Drum Head
3. Conical head: Eg. Jersey Wakefield
4. Savoy type: Eg. Chieftain

Myers classification based on size and shape of head (Allen, 1914)

1. Wakefield and Winning Kadt group-Small, conical, early, Eg. Jersey Wakefield.
2. Copenhagen Market-Round, early, large heads, Eg. Copenhagen Market
3. Flat Head / Drum Head-Eg. Pusa Drum Head
4. Savoy group-Wrinkled leaves, high quality, Eg. Drum Head Savoy
5. Danish Ball head-Thin leaves, Compact head, fine texture, Eg. Danish Ball Head
6. Alpha group-Earliest group, very small head, solid Eg. Miniature marrow
7. Volga group-Thick leaves, loose bottom, Eg. Volga
8. Red cabbage-Similar to Danish ball head but have red colour leaves, Eg. Red Rock.

Breeding in India: Cabbage was introduced much earlier than cauliflower by Portuguese. It was grown during Mughal period. However area increased during British rule. But area under cabbage is smaller as compare to cauliflower.

Prominent Breeding Stations in India

1. IARI, R.S., Katrain (Kullu Valley) H.P
2. Dr. Y.S Parmar University of Horticulture and Forest, Solan, Himachal Pradesh

Objectives

- Development of SI but cross compatible lines for use in hybrids.
- Development of cultivar/ hybrids which can grow under mild winter condition (Tropical).
- Higher yield.
- Varieties resistant to Black rot: (*Xanthomonas campesrtis*) and cabbage yellows (F.O. f sp. Conglutinans., cabbage butterfly, caterpillar, aphids (*Brevicorne brassicae*), Diamond back moth (P*lutella xylostella*).
- Varieties which can stand longer in the field after head formation.
- Narrow, short and soft core.
- Short stem.

Selection Techniques

1. Head shape: Pointed, flat or round heads are preferred depending upon consumers, spherical head ratio is 0.8 to 1, drum head is : 0.6 or less, conical head is : > 1.9 (ratios of polar and equatorial diameter).
2. Heading types and non heading types: Wrapper leaves tight to form head.
3. Medium sized heads preferred.
4. Plant height: short stem is desirable. Tall plant likely to fall.
5. Core width: narrow core is desirable.
6. Core length: < 25% of head diameter is preferred.
7. Core solidity: Soft core is preferred over tough one, especially for processing.
8. Axillary heading: Undesirable.
9. Storability: longer storability is desirable and is positively associated with dry matter content and late maturity.
10. Head compactness: Applying pressure by thumbs, one can make out approximately.
    1. Position of uppermost wrapped leaf: >2/3rd area of head covered - compact
    2. Small core: compactness
    3. Pearson formula: $Z = \frac{C \text{ X } 100}{W^3}$

       Where, C= net weight of head,

       W= average of lateral and polar diameter of the head,

       Z = compact head.

   4. Net weight of head (without stalk and non wrapped leaves) if high then it is more compact.

11. Frame of the plant: small frames are preferred.

Breeding Methods

1. Introduction : Promising in India-Golden Acre, August, Copenhagen Market.
2. Mass selection-Characters which can be improved are maturity (days to 50 per cent heading), stalk length, number of non wrapper leaves, frame or plant spread, shape of head (polar and equatorial diameter-normal or spherical head: 0.8- 1.0 shape index (SI), drum head: SI < 0.5, conical head: SI > 1.0), compactness of head (associated with short sized core), net weight of head (< 750 g), number of marketable heads, bolting and yield.
3. Recurrent selection scheme and other methods.
4. Line breeding,
5. Family breeding.
6. Pedigree method : Selection 8 (Pusa Mukta) is a selection from EC 10109 X EC 24855, resistant to Black rot and is developed at Katrain
7. Bulk population method.
8. Heterosis breeding : F1 hybrids commercially exploited in cabbage. The phenomenon of self incompatibility is utilized to exploit heterosis. Heterosis is observed for earliness, head size, uniformity in head shape and size and yield. Single cross and double cross hybrids have been exploited. Self incompatible female inbred x good op variety was the system predominant till late eighties in USA (1980). There are problems like inbreeding depression, sib seed contamination, reduced incompatibility by environmental condition and restricted pollination within the parental lines by honey bees instead of random mating by bees leading to seed mixture. Depending upon parental lines and condition during seed production sib seed may vary from 0 to 80% (wills et. al 1979). Electrophorosis was carried out on 10 % polyacrylamide gel slabs with gel buffer 0.37 M tris-HCl, pH 9.1 and electrode buffer 0.37 M tris-glycine, pH 8.3. Acid phosphatage activity was found in 3 anodal zones. Assessment is very important for percentage of sib seeds. Seedling markers can be used to quantify the mixture but takes 10 to 12 weeks. Isozyme analysis would be faster technique. Extracts of seeds of F1 were separated by electrophoresis on polyacrylamide gels and stained for 14 enzymes like acid phosphatage, alpha-amylase, alcohol dehydrogenage, carbonic anhydrase, carboxylesterase, catalase, catachol oxidase, betagalactosidase, beta glucosidase, isocitrate lyase, lipase, leucine amino peptidase, peroxidase and pyrophosphatase.

F1 seed production: Use of self incompatibility: Two inbred lines of good specific C.A. for yield are selected which are self incompatible. These two lines are SI and cross compatible because of different S alleles. Single dominant marker gene (eg. purple stem pigment) can be tagged with the male line and have to collect seeds from female only, otherwise (If marker not used) collect seeds from both the parents.

At IARI (Katrain and ND) developed hybrids. Double cross hybrids are produced to overcome problem of inbreeding depression for seed yield.

Synthetics: Set of inbred lines tested for their specific combining ability effects and differing in the S alleles. These inbred lines are grown in alternate rows in isolation. Random mating is allowed and seeds collected from such a random mated population form synthetic seeds. Steps are same as F1 hybrids development. At IARI, Katrain synthetic variety developed by using 7 parents viz., ARU Glory, Pride of India, India Market, Verma's Pride, Pride of Asia, Golden Acre and Pusa Mukta are 7 parents.

Interspecific hybridization carried out to generate variability.

Mutation breeding: Spontaneous mutant characterized by the lack of waxy layer on leaves has been identified. The absence of wax is associated with a significantly higher content of ascorbic acid, dry matter and coarse fiber compared with normal type. Mutant 19P-2 of variety Kjure17 was induced by irradiating seeds with 60 kr gamma rays was semi sterile. Mutation breeding is resorted to eliminate defects and to induce characters like male sterility.

Resistance Breeding

Cabbage Yellows: It is caused by *Fusarium oxysporum*. It is soil borne, vascular wilt favored by warm soil temperatures with optimum at 28°C. There is progressive yellowing followed by brown necrosis, stunted plant growth with premature leaf drop. Type A resistance is determined by one dominant gene and is not influenced by temperature. Type B is conditioned by several genes and breaks down at temperature above 22^0 C. Screening for A type resistance is done by dipping young seedlings in an inoculation suspension and then growing them at 27^0 C. In 2 to 3 weeks susceptible plants will be dead.

Black Rot: It is bacterial disease caused by *Xanthomonas campestris*. Resistance was reported in Early Fuji. It is seed borne, shows vascular bacteriosis causing yellowing of leaves. Resistance is controlled by a major gene 'f' plus 2 modifiers, one dominant and one recessive. For artificial inoculation a bacterial suspension is sprayed on well developed plants early in morning. This introduces bacteria into guttation droplets. In 2 to 3 weeks, susceptible plants will develop large lesions on leaf margins and blackening through veins of leaf and stem. Resistant cultivars will show slight necrotic infection at leaf margins. Symptoms are yellowing of leaves, large lesions, blackening through veins of leaf V shape. Resistant will show necrotic infection at leaf margin.

Complete breeding programme for development of improved disease resistant variety

Line A: Poor quality, resistant to downy mildew (having single dominant gene for resistant to two races)

Line B: Desirable inbred susceptible to downy mildew (DM), resistant to *Fusarium* yellow (FY).

Generation 1 A x B emasculation of B must

Generation 2 20-30 seeds raise F_1 and B line
Screen F_1 for resistance to DM both the races
Emasculate B and go for backcrossing.

Generation 3 Plant 500-600 BC_1 seeds and cultivar B
Screen for resistance to DM, ¼ should have resistance to both the races. BC_1 X B

Generation 4 Sow 500-600 BC_2 seeds in green house, test seedling with fusarium yellow inoculums. Include some seedlings of susceptible character. All seedlings should be resistant to yellow.

Generation 5 BC_2 F_2

Generation 6 BC_2 F_3

Inbreeding and Selection

Trails of advanced lines

Commercial F_1 hybrid seed industry in USA: Ratio of 1: 5 rows if one (female) line is compatible and 2:2 rows if both are incompatible. Seeds harvested from both the parents. Self incompatibility is difficult to manipulate. Considerable amount of selfed seeds observed if used for many generations. Seed lot is discarded if inbred contamination is about 5%.

Salient breeding achievements

Disease	Resistant variety
Tip burn (non-parasitic)	Wisconsin
DM (*Peronospora parasitica*)	January King
Yellow (*Fusariun oxysporum* f. *conglutinaus*)	Wisconsin Ball Head
Club root (*Plasmodiophora brassica*)	Late Moscow
PM (*Mycosphaerella brassicola*)	Louisiana
Black rot	Louisiana and Pusa Mukta)
Mosaic	Badger Ball Head

Indian Cabbage Varieties Belongs to Group 2 (CHM) and 3 (FD or DH)

1. Copen Hagen market group

This group characterized by early, round head with compact heads having few outer leaves and small core. These are more preferred.

Golden Acre: Earliest selection from Copen Hagen market, it takes 60 to 65 days for maturity and bears 1 to 1 ½ Kg heads and it is susceptible to cracking.

- This is an early maturing variety developed by IARI Regional Station, Katrain.
- Plants are small, compact with few outer and clip shaped leaves having short stalk.
- Heads are uniform, very compact, cover leaves dark green, solid and cup shaped with prominent veins; matures on 60-75 days after transplanting; suitable for winter season cultivation in the plains and summer season cultivation (off season) in the hills; gives yield of 20-24 t/ha.
- This is popular among the farmers of J&K, HP, Uttaranchal, Punjab, UP, Bihar, Jharkhand, Chattisgarh, Orissa, AP, MP and Maharashtra. Seeds could be obtained from the branches of NSC.

Pride of India

- This variety is of early maturity group, introduced by YSPUH&F, Solan.
- Heads are small, round, 1-2 kg of weight, outer leaves are few and slightly cup shaped; gives an average yield of 40 t/ha in 60-7 0 days of crop duration.
- Seeds could be obtained from branches of NSC.

Copen Hagen Market: It is later than pride of India and bears 2.5 to 3 Kg heads.

Pusa Mukta (Sel. 8)

- This is a bacterial rot resistant variety developed by IARI Regional Station, Katrain; derived from the cross EC-24855 x EC-10109.
- Plants have short stalk, medium frame and light green wavy puckered at the margin of the leaves.
- Heads are compact, flatish-round with loose wrapper leaf at the top, 1.5-2.0 kg of weight; gives an average yield of 20 t/ha.
- This is popular among the farmers of HP and Uttaranchal.
- Seeds could be obtained from the developing center.

2. Flat Dutch group/ Drum head group

Plants medium to large, outer leaves are large and numerous curving inward enclosing the head loosely, Colour is light green, heads are large, flat and fairly solid, preferred for bulk supply as they have larger heads and often late.

Pusa Drum Head

- This is a late season variety developed by IARI Regional Station, Katrain.
- Heads are large (3-4 kg), flat, cover leaves light green; field resistance to black leg (*Phoma lingum*); ready for picking on 80-90 days after transplanting; gives an average yield of 30 t/ha.
- This is popular among the farmers of Nilgiri hills Chattisgarh, J&K, HP, Uttaranchal and Darjling hills.
- Seeds could be obtained from the developing center.

September: Introduction from Germany, popular Niligiri in Hills, head weight is 3 to 5 Kg and matures in 96 to 110days.

Late Drum Head: It matures in 105 to 110 days,

Early drum Head: It is early type in Drum Head group.

3. Wake Field group

Jersey Wake Field It has pointed head and matures in 55 to 60 days and head weight is 1 to 1.5 Kg and has good for keeping quality.

4. Savoy Group

Not popular in India, Blistered leaves, all shapes of heads, pointed, round and flat Ex: Chieffain.

5. Red cabbage Group

Not popular in India and used for novelty.

Red Acre: It has wax coat and heads weigh 2 to 3 Kg and matures in 90 days and heads are somewhat round in shape.

New varieties

Pusa Agethi: New Heat Tolerant Cabbage: Developed from TKCBH-28 (F1 Hybrid) from Taiwan. IARI New Delhi. 70-90 days, 600-1200g/head.110-to 380q/ha. Flattish-round, compact, non cracking, with large wrapper leaves released during 2000.

F_1 Hybrids: From some seed companies.

Maharashtra Hybrid Seed Company

1. Hari Rani Goal: 1.92 Kg, 95 days from Nursery, 50-56t/ha good keeping quality, resistant to yellows and black rot.

2. Kalyani: Round 1.62 Kg, 94 days from sowing, 50 to 55t/ha
3. Kranti: Round, 1.02Kg (small), 93 days from Transplanting, Tolerant to Cabbage Yellows and Black Rot
4. Sri Ganesh Gol: round, 2 kg, 90-95 days, resistant to cabbage yellows, good keeping quality.

NSC: Marketing two F_1 Hybrids

A. Green Express: Round, Early 2-3 kg

B. Green Boy: Round, Early, 2-3 kg

IAHS: Kaveri : 65-75, 2 kg resistant to heat tolerant, Drum Head

Ganga: Red, 2.5 kg, 80-85 days, 2-3 kg,

Yamuna: Tolerant to Fusarium Wilt race1.

Krishna: Resistant to black rot, 1.5 to 2.5 kg, resistant to non bursting long storage.

19

Cauliflower and Other Cole Crops

Cauliflower follows cabbage in importance with regard to area and production in the world. However in India Cauliflower is more widely grown than cabbage. This crop grows at a latitude 11°N to 60°N with average temperature ranging from 5- 8°C to 25-28°C. It can stand to the temperature as low as -10°C and as high as 40°C for a few days. Italy and India are the major countries growing CF on large scale. Important cauliflower growing states in India are UP, Karnataka, West Bengal, Punjab and Bihar. It is commonly grown in Northern Himalayas and Niligiri hills in south.

Origin and Distribution: Cauliflower is thought to have been domesticated in the Mediterranean region since, the greater range of variability in the wild types of *B.oleracea* is found there (Nieuwhof, 1969). Wild Cole warts are still found in wild state in sea cost of England and Denmark and in the North West France and various localities from Greece to Great Britain. The Cole crops, including cauliflower and cabbage have descended from a common Kale like ancestor, the wild cabbage (*B.olearcea var. sylvestris* L.) which is still found in western and southern Europe and North Africa (i.e, sea coasts of England and Denmark and in North West France and various localities from Greece to Great Britain). Cyprus and area around Mediterranean courts are considered to be the primary centers of origin for Cauliflower. It is also reported that cauliflower is originated from a related species. *B.cretica*. As per Bosewell (1949), it originated in the islands of Cyprus from where it moved to other areas like Syria, Turkey, Egypt, Italy, Spain and North Western Europe. Cornish types originated in England fallowed by temperate types originated in Germany and Netherlands in 18th century. The present Indian tropical cauliflowers developed as a result of intercrossing between European and Cornish type. In India tropical types with resistance to high temp and high rainfall conditions have been developed. Tropical types are grown in Indian plains from May to September, followed by temperate types known as snowball cauliflower.

Crop Improvement works in India are being carried out at IARI, New Delhi, PAU Ludhiana, GBPUAT, HAU, Hissar, and HPAU Palampur. Temperate types improvement work is being carried out at IARI, RS, Katrain and Dr Y.S. PUHF, Solan.

Cytogenetic Studies: It has chromosome number 2n=18. As per the studies on secondary association have been interpreted as showing basic chromosome number x=5, making cauliflower modified amphidiploids from a cross between two primitive 5 chromosome species with subsequent loss of one pair of chromosome. In another study 3 groups of 2 bivalents each and 3 groups of one bivalent each have repeatedly found. By this secondary pairing of chromosomes, it has been concluded that they all derived from a basic species with x=6. The six chromosomes are designated by the letters A,B,C,D,E&F. Later it was indicated that the minimum number of association groups which included all chromosomes was three. This implies that basic chromosome number is three i.e., X=3.

Genetic Studies

Plant height: Monogenic (TT) Tall dominant to dwarfness
Cotyledon size: Monogenic (LcLc) Large dominant to small
Leaves in Curd: Monogenic (LL), Absence of leaf is dominant to presence.
Earliness: Monogenic (EaEa) Earliness dominant to late
Puffiness of bud: Monogenic (phph), non puffy bud dominant to puffy bud
Blindness (absence of curd): Monogenic (bc bc) presence of curd is dominant to absence.
Male sterility: Monogenic, fertility dominant to sterility
Curd colour: Monogenic, white dominant to colour
Leaf colour: (Presence of wax): 'I' suppress biosynthesis of wax and inhibit expression of 'G'.
I- : bluish green leaf colour: glossy iiG-: Normal leaf with wax
Stalk length: 'St' long stalk dominant to short stalk 'st'
Leaf apex: Round: 'Ro', pointed: 'ro'
Plant type: E-Errect, e-flat
Length of flowering stalk: 'F': Long, 'f' : short
Curd colour: Y: Yellow, y: White
Colour of flowering stalk: V: variegated, v: green
Stigma (style) size: Sl: Long, sl: short.

Green curd colour (Crisp and Angel, 1985)

Phenotype	Genotype	Expected ratio in F2
White	Wi-gr^1gr^1	3
Pale green	Wi- gr^1 gr^2 (vary pale)	6
	Wi- gr^2 gr^2 (darker green)	39
Yellow	wiwi gr^1 gr^1(totally yellow)	1
	wiwi gr^1 gr^2(greenish yellow colour)	23
Green	wiwi gr^2 gr^2	1

Dominant genes: Wi-white dominant over yellow, Codominent genes (gr1 gr2) for green leaves. Dominant and epistatic effects contributed more towards the inheritance of curd weight, curd size, plant height, number of leaves and plant spread. Over dominance also played in the expression of these characters. Partial dominance observed for leaf size. In general additive, additive X additive components played more important role for curd maturity (Singh, *et al.*, 1976). Curd weight significantly correlated with curd size, leaf size and plant spread at phenotypic and genotypic level.

Self incompatibility: For clear cut distinguishing between incompatibility and compatibility carry pollination at 14 to 15°C or at 10 to 12°C and subsequently style fixed after 20-24 hr and 40-48 hr of pollination respectively. No commercial hybrid has yet been released from Public sector institute because SI status not studied thoroughly.

Variability and classification of cauliflower

Type	Origin	Probable time of Cultivation	Characters
Cornish	England	Early 19th century	Plants vigorous, long stalked, leaves loosely arranged, broadly wavy, curds flat, irregular loose, not protected, yellow, highly flavored.
Northerns	England	19th century	Leaves petiolate, broad, very wavy, serrated, curds good, well protected.
Roscoff	France	19th century	Plants short, leaveslong, erect, slightly wavy with pointed top, midrib prominent, bluish green curds white or creamy, hemispherical well protected.
Angers	France	19th century	Leaves very wavy, serrated, grayish green, curds solid, white.

Major Types of Cauliflowers

European Cauliflowers: Systematic and extensive cultivation of cauliflower was first started in Italy where the 'originals' were developed. These original Italian types were taken to France, England, Germany and Netherlands where some important local types were developed from them e.g., The 'Northers' in Yarkshire and Darbyshire, The 'Cornish' in Cornwall, The 'Angers' & "Roscoff" in Britanny and the 'Erfurt' or its allied 'Snowball' in Germany and Netherlands.

Indian Cauliflowers: Indian cauliflowers are characteristically different from the types grown in Europe, as they are tolerant to high temperature and humid condition and are perhaps the earliest maturing types known. Indian cauliflowers are dwarf selections of Erfurt or Snowball types. This view is also supported by Nieuwhof (1969) who stated that selections from Erfurt types have yielded early varieties that performed better in warmer regions producing good curds at temperature above 20°C (Some of these varieties are Early Patna, Early Banaras, Early Market). However, Swarup and Chatterjee (1972) questioned this. According to Swarup and Chatterjee (1972), Indian hot weather types have not originated from Erfurt or Snowball types as suggested by Giles or Nieuwhof. Their argument is based on following evidences: 1) Typical Indian cauliflowers belong to maturity group I and IV (September-early November, mid November-early December). 2) In contrast, group III (mid December-mid January) compares favourably with late Snowball or Erfurt types in regard to climatic requirements. From these observations it appears that parental varieties contributing most to the Indian types tolerant to high temperature and rainfall conditions are the Cornish which are predominant types in Group I and II but almost absent in group III. Morphological and other characteristics of long stalk, open habit, exposed, yellow uneven curds which loosen up easily and strong flavour are some attributes common to both Cornish and Indian cauliflower particularly early types maturing in September to November. The observations on self incompatibility also support the view that Indian types are genetically different from Erfurt because of difference in self-incompatibility occurrence. However, in maturity group III some self-compatibility has been recorded. This indicates that Indian types particularly Group I and II have been developed from winter types and not from summer types (Snowball). Cornish, Roscoff and Northerns are classified as winter types and have been reported to be highly self-incompatible. Genes for resistance to black rot are present only in Group I (except one inbred line in group II) while all the varieties tested in Snowball groups are susceptible to black rot.

Indian types	European types
Tolerant to heat	Not tolerant
Curd formation at and above 20°C	Curd formation at 5-20°C
Annual	Biennial
Yellow curds, loose with strong flavour	Snow-white curds with very mild or no flavour
Early	Late
More variable	Less variable
More self-incompatible	Less self-incompatible
Short juvenile phase	Long juvenile phase
No need of vernalization but needs cold treatment at 10-13°C for 6 weeks	Needs vernalization at 7°C for 8-10 weeks

Different European Cauliflowers

1. Originals or Italian: Mediterranean origin
2. Cornish : England origin
3. Northerns: England origin

4. Roscoffs: France origin
5. Angers : France origin
6. Erfurt and snow ball: Germany and Netherlands

Pollination Techniques

Selfing: varieties can be slefed by simply bagging the flower stalk or by caging some flies insides with self incompatible plants, bud pollination gives better results. In incompatible lines the pollination is carried out in buds before 2 to 4 days of opening with emasculation or without emasculation.

Crossing: Flowers may be emasculated by removing 6 stamens using a pair of forceps. In self compatible lines (European types) the stamens are removed before opening of the buds as the flowers already fertile in bud stage, crossing can be done at the same time. In SI types, emasculation may be omitted. Put the crossable plants in cages with honey bees or bumble bees and flies.

Character Variability and Ideotype

Plant Types: There are four distant plant types. Type 1 with Completly flat leaves with exposed curds. Type 2 & 3 are intermediate, 2 being close to 1 and 3 being close to 4. Type 4 completly erect leaves with small curd. Plant type 3 is the best as it has long erect leaves with or without self blanching habit and has medium sized curds. A lower plant type is generally dominant over the higher (erect).

Stem Pigmentation: Generally the stem of Indian cauliflower seedling is green unlike Snowball which is pigmented. However some Indian cauliflower, stem pigment is quite pronounced, almost like Snowball. If it is dominant character it will be useful as a marker gene.

Stem Length: Stem length can be Short 15cm, Medium 16 to 20cm, and Long 21cms and above (Indian CF var Sept. to Dec.). Group I and II is medium while Group III is short (Dec to Jan maturity). Short stem is dominant over long. Generally shorter stem length is preferred.

Leaves: Long and narrow types available in addition to long and broad, short and broad types. Margins are straight, broadly wavy or sinuate. Colour varies from bluish green to waxy green or glossy green.

Curd: Weight varies from 200g to 1kg or more, early types have small size and late types are having big size curd. Curd shape varies with Flat, semispherical or slightly conical. Compact curd is preferred and curd weight is 35 to 40 per cent of total plant weight. Quality of the curd is determined by the compactness and colour. Colour varies from white (snowball) which is best to yellow to creamish white which is observed in Indian cauliflower. Defects is curds like riceyness.

Grades like 1 to 4 available where grade 1 is the best and perfect and no female buds on the surface of the curd and grade 4 you will find well formed buds on the surface. Ricey curd is poor quality for market, but set more seeds.

Ideotype: (Chatargee and swarup, 1972): Ideal types have been designed and accordingly objectives can be set. Maturity: Specific group (75-80% uniformity), Stalk length: 12-15cm, Plant type: No.3, Frame: 35-45cm, Leaf number: 18-22, Leaf length: 50-55cm, Curd shape: Hemispherical, Curd size (dimension):15-18cm, Curd weight: 750-1Kg, Curd colour: Retensive white and variety should be resistant to black rot and stump rot.

Objectives

1. High yield and earliness
2. Good quality curd: White, compact, free from malformation, curd weight <750g.
3. Non bolting, erect leaf orientation, self blanching.
4. Varieties of tropical types where curd form in Aug-sept.
5. Varieties which can form curd during summer and rainy season in hills to feed rest of country.
6. Varieties with good seed set and seed yield.
7. Exploitation of Heterosis: SI but cross compatible inbred lines with good combining ability.
8. Resistance to diseases: Black rot, *Sclerotia* rot, *Alternaria* blight, Downy mildew, cabbage yellow, *Erwinia* rots.

Breeding Methods: Mass selection, pedigree method, heterosis breeding and bulk population methods were followed and also suggested.

Heterosis Breeding: Inbreds of self incompatible lines are produced through bud pollination, late pollination, electrically stimulated pollination, thermally stimulated pollination, stigma mutilation and other methods. Inbreeding depression in cauliflower is not very high compared to other cole crops. Self incompatibility lines can also be maintained through clonal propagation also. Single cross hybrids and double cross hybrids and synthetics have been developed in cauliflower.

Steps

Single cross and double cross hybrids

1. Inbred genetic divergence study and grouping.
2. Selection of lines based on genetic divergence and desirable plant characters.

3. Diallele crossing.
4. Combining ability assessment
5. Identification of heterotiotic F1 hybrids.

Synthetics

1. Development of inbred lines
2. Testing for sca
3. Selection of 5 or more lines whose combination have maximum sca effects.
4. Growing inbred lines in isolation and random mating.
5. Test the performance of synthetic over commercial variety.

   Ex. Pusa Early synthetic: 25°C curding temperature

Pusa synthetic: 10-15°C curding temperature.

Interspecific Hybridization: Interspecific hybridization can be resorted to generate variability.

	B.nigra	*B.oleracea*	*B.campestris*
B. nigra (x=8)	SI & SC	CC	CC
B. oleracea (x=9)	CC	SI & SC	CC
B. campestris (x=10)	CC	CC	SI & SC

Intergeneric Crosses: Cauliflower and radish was fairly successful only when cauliflower acts as male parent. Failure of reciprocal crosses was due to inability of radish pollen to penetrate in to the styles of the cauliflower.

Breeding for Curd Quality

Selection Against Bracting Defect: The cauliflower curd consists of a mass of compressed, branched peduncles bearing many thousands of pre-floral meristems. White bracts corresponding to the auxiliary leaves are usually present inside the curd. Often these bracts grow through the surface of the curd and give it a papillae appearance which reduces its commercial value. Its value is further reduced if the bracts develop green or purple coloration. Normally assessment of bracting is made microscopically in the field and efficiency of selection is correlated with the stringency of this assessment. It is therefore, necessary to develop techniques which maximize the expression of the character without affecting the assessment of other characters. Crisp *et al.* (1975) reported that when curds were taken from field grown plants and aspetically cultured, bracts had usually started growing after about 20 days in culture and their relative sizes were recorded at this stage as small (score 1) to very large (score 4).

Breeding Against Ricyness: Precocious flower bud formation i.e., ricyness is appearance of outgrowth of about 1mm diameter from curd surface. These structures are immature flower buds. This character is relatively consistent and is induced by cold temperatures. That is why cauliflower varieties bred in tropical produce ricy curds if grown under cool conditions. However, the reverse is not true. This defect has been shown to be highly heritable, monofactorial and hence highly responsive to selections (Crisp and Tapsell, 1993).

Breeding for Optimum Leaf Geometry: Variation exists among cauliflower accessions in germplasm for geometric configuration of leaves. Most of cauliflower produced in USA is hand-tied to prevent head discolouration due to sunlight exposure. Discoloured or yellow heads of cauliflower fetch lower price. Therefore, cauliflower varieties should be developed in which the leaves assume a more upright position around the developing head, rather than a horizontal position. The upright leaves provide partial covering around the head and protect it from sunlight and eliminate the need for tying. Chatterjee and Swarup (1972) have described four plant types and recommended Plant Type No. 3 which corresponds to plant type of Snowball. Plant Type 3 is considered the best as it has long erect leaves with or without the self blanching habit and has medium sized curds. Plant Type 1 has completely horizontal leaves and Plant Type 4 has erect leaves, whereas Plant Type 2 and 3 are intermediates. Type 2 is being closer to Type 1 and Type 3 is being closer to Type 4. The inheritance of leaf geometry was studied by Werner and Honma (1980) who evaluated P1, P2, F1, F2, BC2 of MSU 839 (upright) X MSU 831 (horizontal). A relative measure of leaf geometry for an individual leaf was determined in the manner.

Genetics of Leaf Orientation

Generation	Upright	Horizontal	Fit for Ratio
F2	330	73	54:10
BC1	28	34	1:1
BC2	76	6	1:0

A-B-C-, A-B-cc, aaB-C-, A-bbC-: Upright, aabbC-, aaB-cc, A-bbcc, aabbcc: Horrizontal

At least two genes in dominant conditions lead to upright. F1 have advantage of both the parents.

Breeding for Resistance to Black Rot: An important work on this aspect was carried out by Sharma *et al.* (1972) in India. They evaluated parental lines, BC_1, BC_2 and F_2 generations from five crosses involving MGS (highly resistant), Pua Kea and S. No. 445 (resistant) and S. No. 246, S. No. 15, EC 12013 and

EC 12012 were susceptible. Resistance is governed by quantitative genes with dominance (Sharma *et al.*, 1972). Screening can be done by soaking the seeds in liquid inoculums for 24 hr and spraying thrice at 8 days intervals at seedling stage with inoculums and adult plants can be twice inoculated by spraying and cutting the leaf tip inside inoculum. Resistance to black rot is polygenic dominant. Dominance component was greater than additive component in majority of crosses. It is suggested to attempt backcross breeding or selection in segregating generation of resistant x resistant cross to evolve highly resistant lines. Heterosis breeding involving both the parents resistant in single cross may be useful in obtaining F1 hybrids with high level of resistance.

Breeding Achievements

Group I: Mid September to mid October maturity group: Pusa Katki and Early Kunwari

Group II: Mid October to mid November maturity group: Pusa Shubra, Synthetic I, Pusa Deepali

Group III: December maturity group: Hissar I, Pusa synthetic and Pusa Shubra.

Group IV: It is late maturity group or snow ball group: Snow ball 16, Pusa snow ball, Pusa snow ball 2

Resistant to whiptail: Lecerf, M16/1

Resistant to Mosaic: Native Dwarf Erfurt, AH46

Resistant to *Rhizoctonia* rot: South specific

Resistant to Striped flea beetle: Snow ball A

Cold Tolerant: Svale, Fanolz

For specific temperature: Curding at 25°C: Pusa Early Synthetic

20-25°C: Pusa Deepali,

16-20°C: 11-C,

10-15°C: Pusa Synthetic.

Arka Kanthi: It was developed from IIHR, Benagaluru

Early Types

Early Kunwari: Developed and released from PAU, curd semispherical with even surface.

Pant Gobi-3: It is synthetic variety (8 Inbreds of Katki group), long stem, non ricey curds, 110 days for maturity and yields 120q/ha and released from GBPAUT, Panthnagar.

Pusa Katki: Released from IARI and it belongs to October- November maturity group, medium plants and it is early (sowing in middle of May).

Improved Japanese: Introduction from Israel, does not tolerate to hot season, it has compact white curds and optimum temperature requirement is 16 to 20°c.

Pusa Deepali: This variety has been developed by IARI, New Delhi. Plants are erect with short, waxy, green leaves (rounded tips) and self blanched. Curds are compact, white, ready for harvest by mid October; suitable for planting from early to mid July in northern plains and gives an average yield of 15 t/ha. This is popular among the farmers of Punjab, UP, Bihar and Jharkhand. Seeds of this variety could be obtained from Division of Vegetable Crops, IARI, New Delhi. It belongs to October-November maturity group. It has white medium sized curds, no riceyness, sowing in end of the May to early June.

Pusa Early Synthetic: Developed by IARI, New Delhi (1990). Plants are erect with bluish green leaves. Curds are small-medium, flat, creamy white and compact, it is suitable for early planting and rresistant to riceyness. It gives an average yield of 11.7 t/ha.

Medium Season Varieties

Pusa synthetic: It involves 7 inbred lines with good gca and developed at IARI. It has creamy white compact curds and it belongs to mid Dec to Mid Jan (12-15°) maturity group. Planting season is mid September to late September in North India.

Pant Subra: Developed through simple recurrent selection and released in 1985, curds compact, slightly conical, creamish white in colour, non riceyness curds and free from leafy curd. Adopted for rainy season and developed at GBPUAT.

Hissar 1 : It has medium to large sized heads with 25t/ha yielding ability.

Punjab Giant 26: It has solid curds, medium in size and released from PAU Ludhiana.

Pusa Subhra: This variety has been developed at IARI, New Delhi. Plants are erect with long stalk and bluish green leaves. Curds are white, compact, 700 to 800 g of average weight and free from riceyness. It is resistant to black rot and gives an average yield of 20 t/ha in 90-95 days. Seeds could be obtained from Division of Vegetable Crops, IARI, New Delhi. For curding optimum temperature is 12 to 16°c. It is suited for August to September sowing.

Late Varieties

Pusa Snowball 1: It is the derivative of cross between EC12013 and EC12012, released during 1977. It is a late variety suitable for cool season. Optimum temperature for curd initiation and development is from 10 to 16°C. Sowing time in NI is mid September to end of October. Inner leaves slightly cover the curd. Curds are very compact, medium size and snow white. It is developed at Katrain, IARI regional station.

Pusa Snowball 2: Released during 1972 from Katrain, curds remain white even at exposure to light. Cultivar could not become popular because of its poor seeding ability.

Pusa Snowball Kt-25: This late maturing variety has been developed at IARI Regional Station, Katrain. Leaves are waxy, upright, slight bending towards inner side with puckered margins. Curds are very solid, medium sized, white with good keeping quality. It is suitable for transplanting from October to early November. It is tolerant to sclerotia rot and black rot diseases and gives yield of 20 to 30 t/ha.

Pusa Snowball K-1: This is a late maturing variety developed at IARI Regional Station, Katrain. It is derived from an exotic collection. Leaves are slightly puckered and wavy. Curds are self blanched, snow-white and compact. It is moderately resistance to black rot under field conditions. It matures in 90 to 95 days after transplanting and gives an average yield of 22.5 t/ha. This is popular among the farmers of J&K, HP, Uttaranchal, West Bengal and Assam. Seeds could be obtained from the branches of NSC. Amongst the snowball types it has best quality curds. It is late than Pusa Snow ball 1 & 2 and extends the availability of cauliflowers in market.

Other Varieties

Dania: It is developed from IARI, Regional Station, Kalimpong for eastern hilly areas. Plants are strong having medium sized curd. This variety is tolerant to the stress conditions.

Pant Gobhi-2: It is composite variety developed from GBPUA-T, Pantnagar. Curds are ready for harvesting from October onwards. An average yield of this variety is 100 q/ha.

Pant Gobhi-4: It is released from GBPUA-T, Pantnagar by simple recurrent selection of Aghani group and recommended for cultivation in 1995 for Uttar Pradesh state. Plants are medium in growth with upright leaves. Curds are round, creamy white and solid. This variety is free from riceyness.

Punjab Giant-35: It is suitable for late season cultivation. Curds are compact, medium sized and snow white. The average yield of this variety is 225 q/ha.

Pusa Aghani: This variety has been developed from IARI, New Delhi. Curds are ready for harvesting in the month of November-December. Curds are big in size, solid and white in colour. Yielding ability of this variety is 150 to 160 q/ha.

Pusa Himjyoti: It is released from IARI, New Delhi and suitable for transplanting in hill tract in the month of May and August. Plant is straight with bluish green leaves. Curd is quietly white, solid and round. It will be ready for harvesting in 30 days after transplanting. Yield of this variety is 160 q/ha.

Pusa Sharad: It is released from IARI during 1999. Plants are semi-erect and open types with small stalk. Its marketable curd becomes ready 85 days after transplanting. It is two weeks early to standard cultivar, 'Improved Japanese.' Curds are white, knobby, very compact, semi-dome shaped and about 750 to 1000 gm weight. Average yield is 260 q /ha

Kashi Agahani: Curd is white, compact, medium-sized, slightly dome-shaped, weighing about 850 to 900g. The plant is semi-erect type having medium growth habit with medium-sized stalk. Leaves are waxy, bluish green. Inner leaves not covering the curd initially. It takes about 70 to 78 days from transplanting to first harvesting. It is moderately resistant to downy mildew and can tolerate high temperature and rainfall.

Hybrids

Pusa Hybrid-2: It has been developed by IARI, New Delhi. Plants are semi-erect with bluish-green upright leaves, curds are creamy-white, highly compact, average weight 907g. It matures from mid-November to mid-December in the North Indian Plains. It is resistant to downy mildew and gives an average yield of 25 t/ha. This is popular among the farmers of Punjab, UP, Bihar, Jharkhand, Chattisgarh, Orissa and AP. Seeds of hybrids and parental lines could be obtained from the Division of Vegetable Crops, IARI, New Delhi.

Punam: This hybrid is of 95 days maturity group, developed by Beejo Sheetal Seeds Pvt. Ltd., Jalna. Curds are solid with 1.5 to 2.0 kg of average weight and gives yield of 20 to 25 t/ha. This is popular among the farmers of J&K, HP and Uttaranchal. Seeds could be obtained from the developing center.

Priya: This hybrid is of mid-early maturity group, developed by Beejo Sheetal Seeds Pvt. Ltd., Jalna. Curds are solid, milky white with 1.0 kg of average weight and protected by dark green outer leaves and gives yield of 20 to 25 t/ha. This is popular among the farmers of J&K, HP and Uttaranchal.

Ageti Himlata: This is an early maturing hybrid developed by Century Seeds Pvt. Ltd., New Delhi. Curds are very compact, medium sized, firm, dome shaped with smooth texture. It is able to retain pure white colour even in hot sun. It is suitable for early spring sowing in western and southern region and early summer sowing in North India. Seeds could be obtained from the developing center.

Private seed companies are marketing F_1 hybrids seeds.

SEMINIS: Sukanya, Subhra, Megha

NS-60

This popular hybrid with mid season maturity group has been developed by Namdhari Seeds Pvt. Ltd., Bangalore. Curds are attractive, compact, domc shaped, milky white with an weight of 1.25-1.5 kg and good firmness; gives an average yield of 25 t/ha in 60-65 days of crop duration (May-Sept in South India). This is popular among the farmers of J&K, HP, Uttaranchal, Punjab, UP, Bihar, Jharkhand AP, MP, Maharashtra, Karnataka, Tamil Nadu and Goa. Seeds could be obtained from the developing centre.

NS-66

This hybrid with mid-late season maturity group has been developed by Namdhari Seeds Pvt. Ltd., Bangalore. Curds are attractive, compact, smooth, domc shaped, pure white with an averager weight of 1.5-1.7 kg and good firmness; gives an average yield of 25 t/ha in 70-75 days of maturity duration (Sept-Oct in North India). This is popular among the farmers of J&K, HP, Uttaranchal, Punjab, UP, Bihar, Jharkhand AP, MP, Maharashtra, Karnataka, Tamil Nadu and Goa. Seeds could be obtained from the developing centre.

NUNHEMS: Himpushpa, Shweta

IAHS: Swati, Himani

Summer King: This hybrid has been developed by Sungro Seeds Ltd., New Delhi. Curds are creamy white, dome shaped, compact with weight of 400-600 g and self blanching; matures at about 55 days after transplanting;suitable for summer season cultivation. This is popular among the farmers of UP, Bihar, Haryana, Punjab and Rajasthan.

Knol- Kohl: *Brassica oleraceae* var. ***gongiloides***

It is not widely grown in India except Kashmir, West Bengal and some Southern states. It is characterized by formation of knob which arises from the thickening of the stem tissue above the cotyledon. It is the portion which is used as a food although young leaves are also cooked. Because of limited importance of the crop there are hardly centers working for its improvement. Commonly grown varieties are introduction and being maintained and multiplied. Most common once are White Vienna, Purple Vienna (within these there is variation and strains are there).

White Vienna: It is an early variety. Plants are dwarf with short tapped foliage and medium green. The knobs are globular round, flesh tender and crisp. It takes about 50 to 55 days for knob formation after transplanting. An average yield of this variety is 150 q/ha.

Purple Vienna: It is a late group variety having purple colour leaves and stem. Knobs are big in size with purple colour spot. Knobs are ready for harvesting in

55 to 60 days after transplanting. An average yield of this variety is 150 to 200 q/ha.

Early Purple Vienna: Leaves are purplish in colour. The knobs are globular round large, purple skin with light green flesh. This variety takes 55 to 60 days for knob formation.

Early White Vienna: It is an early variety. Plants are dwarf, short tapped and medium green foliage. The knobs are globular and round. Flesh is tender and crisp. It takes about 50 to 55 days for knob formation after transplanting.

King of North: This variety possess plant height of 20 to 30cm, dark green foliage, flattish round knobs, large leaf sheath which well spread over the knob, broad leaves and takes about 60 to 65 days to knob formation.

Large Green: It is late variety with vigorous growth and dark green foliage. It has flattish round and dark green knobs.

Broccoli *(Brassica oleracea* var. *italica)*

Green Head: It is developed and released from HPKV, Palampur.

Italian Green: It is developed and released from HPKV, Palampur.

Palam Smridhi: It is developed and released from HPKV, Palampur. Each terminal head weigh 400g. Average yield is 50 tones/ha and ready for harvest in 85 to 90 days after transplanting.

Pusa Broccoli (KTS-1): It has been developed and released from the IARI, Regional Station Katrain during 1996. Plants are medium-tall (65-70 cm long). The main head size and weight are about 6.0cm and 15.4cm (height x diameter) and 350 to 450 g. It is ready for harvesting after 125 to 140 days after sowing and 90 to 105 days after transplanting. It is medium maturing variety (Verma and Sharma, 1998).

Punjab Broccoli-I: Identified as variety in 1998 from PAU, Ludhiana. Sprouts are compact, attractive, succulents and suitable for salad and cooking purposes. The yield potential is 175 q/ha.

Aishwaraya: It is a private hybrid from Nunhems Seeds Pvt Ltd. It has compact curds with bright green in colour and average weight is 500 to 700gm.

Fiesta: It is from Bejo Sheetal Seeds Pvt Ltd. Early variety and head weight is 900gm and it is resistant to *fusarium* wilt.

Brussels Spourts (*Brassica oleracea* var. *gemmifera*)

Hilds Ideal: This is an introduction and recommended by IARI, Regional Station, Katrain, Kullu Valley (Sharma and Gill, 1986). Its plant height is 60 to 65 cm and bears 45 to 55 sprouts / plant and number of leaves varies from 45 to 55. Average diameter of sprouts varies from 7.0 to 8.0cm and each sprout weight is about 7 to 8g. Sprouts are compact with good flavour. It takes about 115 days for first picking. Yield per plant varies from 250 to 400g.

Jade Cross: This hybrid matures in 90 days. Sprouts are firm, dark green, closely packed on long stems, can be grown under wide range of growing conditions.

Brussels Dwarf: Plants are dwarf with 50 cm in height, early and suitable for areas of short growing season. Sports are medium sized and high yielding.

Other varieties: Royal Marvel, Rasmunda, Sonara, Ladosa, Olivir, Rogar, Rider, Boxer, Richard, Stephen, etc.

20

Amaranthus

Amaranthus is one of the main genera of large and taxonomically diverse group of leafy vegetables. Amaranthus is the cheapest dark green leafy vegetable because of low production costs and high yield. It is often described as poor man's vegetable.

Origin and Distribution: The main vegetable type *Amaranthus tricolor* L. originated in South or South East Asia, particularly in India. Budhist monks and Muslim invaders took the crop to neighboring countries. Another vegetable type *A. dubius* shows diversity in Central America, Indonesia, India and in Africa. *A lividus* is a popular leafy vegetable in Southern and Central Europe. *Amaranthus* included 50 to 60 species.

Six vegetable species

1. *A. tricolor* L : 2n = 34,
2. *A. blitum* L. 2n =34
3. *A. spinosus* L.-2n = 34, weedy species
4. *A. dubius* Mart.ex Thell. 2n=64 allopolyploid
5. *A. tristis*
6. *Amaranthus lividus*

Weedy species: *Amaranthus spinosus, Amaranthus viridis* 2n=34,

Grain types

1. *A. hybridus :* 2n=32
2. *A. caudatus :* Its origin is Andes of South America
3. *A. hypochondriacus:* Its origin is Asia and Africa
4. *A. cruentus:* Its origin is Gautemala
5. *A. edulis :* It is an autotetraploid and originated in Argentina

A. tricolor, A. dubius and *A. Cruentus* are cross incompatible and there are variants for chromosome numbers.

Amaranthus Sub genera Blitopsis: leafy types
Amaranthus: grain types

Blitopsis	Amaranthus
1. Flower clusters in axils	terminals
2. Green leaf types	grain types
3. Self pollinated	cross pollinated (16-35%)
4. 10 to 25 male flowers per glomurule	0.5 to 1 male flower per glomurule (dichasial cyme) (dichasial cyme)

Genetics: Red plant pigmentation (RR) is dominant to green (rr)
Spininess is dominant to non-spiny.

Objectives

1. High leaf yield
2. Multicut types (day length as flower induction factor influences quantity and quality)
3. High quality-low anti-nutritional factors (nitrates and oxalates), high carotene, iron, calcium, ascorbic acid, folic acid and proteins content.

Breeding Methods

Mass Selection: It is a highly cross pollinated crop (unisexuality) and mass selection is a common method of breeding. Varieties developed through mass selection are:

1. CO2 is a product of local material – *A. tricolor*
2. CO1 – *A dubius* : Green leaves
3. CO3 – *A tricolor* Var. *Tristis* : Green leaves
4. CO4 – *A. hypochondriacus* : Grain type

Pure line selection: Arka Suguna variety developed through pure line selection.

Bulk Population Method: This method can be followed.

Interspecific Hybridization: To generate variability and to transfer desirable characters to cultivated vegetable types interspecific hybridization can be followed. Crossing among species of same chromosome number is possible.

Cross Compatibility Reported

1. *A. cruentus* (2n=32) x *A. hypochndiacus* - Incompatible
2. *A. edulis* (tetraploid) x *A. cruentus* (2n=32) - Incompatible
3. *A. edulius* x *A. hybridus* (2n=32) : Seedlings die
4. *A. caudatus* x *A. hybridus* (2n=32) : Seedlings die

Polyploid breeding: *A dubius* is a natural polyploid (allopolyploid). *A. spinosus* (2n=34) and *A. hybridus* (2n=32) have contributed for *A dubius* (2n=64) development. Colchicine induced tetraploids of *A. blitum* (2n=34) have been obtained which had larger and numerous leaves with long vegetable cycle. *A blitum* value is increased as vegetable. Seeds and seedlings of *A dubius* (2n=64) were treated with colchicine. Polyploid resulted in the reduction of plant height, leaf size and length and caused pollen and seed sterility. That means you cannot increase ploidy level beyond tetraploid.

Salient Breeding Achievements

Klaroen : Tolerant to wet rot (*Choanephora* sp.)

IARI Bred Varieties

Chotichaulai (*A. blitum*), Badichaulai (*A. tricolor*).

Pusa Lal Chaulai, Pusa Kirti and Pusa Kiran are rich in iron, calcium, vitamin A and C.

A43 : It has low nitrate (0.56%) and low oxalate (0.94%) which are anti-nutritional factors. Oxalates bind essential nutrients and create kidney stones.

Kannar Local : Released from KAU and it is short day type (*A. tricolor*).

CO-1 (*A. dubius*): Developed for tender greens and matured stem by TNAU in 1968. Yields 7.165 t/ha at 25th day after sowing, leaf stem ratio is 2.0. It is suited for late harvest. Leaves are dark green with ridged appearance. Stem is dark green, round and succulent, inflorescence terminal and axillary, takes 50 days for flowering and 90 days for seed maturity. Plants are tall (150 cm) and seeds are black.

CO-2 (*A. tricolor*) (1976): Suited for early harvest, it yields 10.78 t/ha at 25th day after sowing, leaf stem ratio is 1.8. Leaves are green and lanceolate and slightly elongated. Stem is green and succulent, spikes terminal and axillary, takes 45 days for flowering and 80 days for seed maturity. Seeds are black.

CO-3 (*A. tristis*) (1981): It is suited for clipping of tender greens, locally known as Araikeerai in Tamil Nadu. Yields 30.716 t/ha green in 10 clippings commencing from 20 days to 3 months. The leaf stem ratio is high. This flowers in 35 to 40 days after sowing and seeds mature in 85 to 90 days.

Arka Arunima : It is a multicut variety having dark purple leaves which are broad, first picking starts 30 days after sowing, 2 subsequent cuttings after 10-12 days interval, total yield of 27.4 t/ha.

Choti Chaulai (*A. blitum*) (IARI): Suited for leafy shoots, plants erect with thin stem, slightly dwarf, leaves small, green in colour, responds well to cutting, suited for early summer and rainy season.

Badi Chaulai (*A. tricolor*) (IARI): Suited for leafy shoots, stem thick, green, leaves large and green, responds to cutting and distinguishable by its longer growing period and suited for summer season.

Pusa Lal Chaulai: High yielding, red pigmented, yields 49 and 45 t/ha of leaf yield with summer - spring and kharif seasons, respectively. Red dye extracted from leaves can be used as natural food additive.

Amont (*A. cruentus*): Developed in Montana, USA from line RRC-A362. It has central main panicle with thick, erect to dropping, finger like branches. Mature in 122 to 127 days. It is 198 cm tall under irrigated conditions and showed no sign of lodging.

Arka Suguna (IIHR-47) (*Amaranthus tricolor* L.): It is a pure line selection from an exotic collection from Taiwan (IIHR 13560). It has light green, succulent stem and broad leaves. It comes to harvesting in 25 to 30 days after sowing and gives 5 to 6 cuts in 90 days. It is moderately resistant to white rust under field conditions and yields 25 to 30 t/ha.

Arka Samraksha: It is a high yielding amaranth variety, with high antioxidant activity of 499 mg (AEAC units) and minimum nitrate content of 27.3 mg and 1.34g of oxalates per 100g fresh weight of leaves. It is a pulling type amaranthus variety with green leaves and stem, yields 10.9t/ha in 30-35 days duration.

Arka Varna: It is a high yielding amaranthus variety, with high antioxidant activity of 417 mg (AEAC units), nitrate content of 37.6 mg and 1.42 g of oxalates per 100g fresh weight of leaves. It is a pulling type amaranth variety with green leaves and pink stem, yields 10.6 t/ha in 30-35 days duration.

Arun (S 8): It is a high yielding (20.1 t/ha) amaranthus variety developed at KAU has attractive, maroon red leaves with average leaf length of 25 cm and average leaf width of 11 cm.

21

Carrot

Carrot (*Daucus carota* L.) is a cool season root vegetable and temperature is the important factor affecting root shape and colour, besides soil type and moisture content of soil. It provides an excellent source of vitamin A and fiber in the diet. The genus *Daucus*, which include carrot, has many wild forms that grow mostly in the Mediterranean region and South West Asia. Representatives are found in Africa, Australia, & North America. For *Daucus carrot* L. it is generally agreed that Afghanistan is the primary center of genetic diversity and therefore the primary source for dissemination. Asia Minor, Afghanistan, North West India, Iran and Turkey are the centers of origin of carrot. Banga (1957& 1963) has provided evidence that the purple (anthocyanin) carrot together with a yellow variant spread from Afghanistan to Mediterranean area as early as 10th and 11th century. It is known in Western Europe in 14th and 15th century, China in 14th century and Japan in 17th century. The European carrot (Carotene carrot) derived from the Asiatic (anthocyonin containing) form of carrot (tropical carrot). Afghanistan is the centre of diversity of the purple coloured carrot (anthocyanin). Integration of Asiatic and European carrots has given the present day forms. The yellow coloured (carotene rich) forms were also thought as mutants of anthocyanin carrots. The yellow mutants were traced in Asia Minor in 10 and 11th centuries, Spain in 12th century, North Western Europe in 14th century and England by 15th century. The crop was introduced to China during 14th century and to Japan during 17th century. *D. carota* and *D. sativus* are only cultivated species of the genera with Chromosome No. 2n = 18. All forms have same chromosome number.

The white and orange (carotene) carrots are probably mutations of the yellow form. Orange carrots were first cultivated in Netherlands and adjacent areas probably in 17th century. The domesticated carrot is closely related to and readily crosses with the highly diverse and widely adopted wild carrot known as 'Queen Anne's Lace'. Naturally Wild accessions moved from semitropical to temperate zones will generally behave as annuals, with bolting following prominently after a brief exposure of seedlings to low temp & longer photoperiod of Northern latitudes. Domestic cultivars have been selected for non-bolting and therefore behave as biennials or winter annuals. Under vernalization they will promptly bolt and mature as seed crop in total of 12 to 13 months.

Daucus

1. *Eucarota:* Annual and biennials: *D. carota, D. maritimus, D. major, D. sativus*
2. *Gummiferea:* Perinnials: *D. gummiferea, etc.*

Classification

Asiatic carrots: Tropical carrot, Eastern carrot, anthocyanin carrot, heat tolerant, deep red or purple coloured, high yielding, and low carotene content.

European carrots: Western carrot, temperate carrot, carotene carrot, cold tolerant, orange coloured and rich in carotene.

Floral Biology and General Botany: Carrot umbel is a compound inflorescence. Primary umbel contains over 1000 flowers at maturity, where as secondary, tertiary and quaternary umbels bear successively fewer flowers. Usually several floral stalks develop form a single plant.

Flowers: Flowers arranged in umbellets and umbellets into an umbel. Within umbellets and umbels, floral development is centripetal and arrangement is spiral. Thus first mature flowers are those on the outer edges of the outer umbellets. Primary umbel consists mainly of bisexual flowers and male flowers can occur frequently in subsequent umbels. The pollen of central flowers of an umbellet are larger and more frequently fertile than that from peripheral flowers. Carrots are protandrous. It has splits-style, which separates when the flower is receptive for pollination.

Petals of petaloid (not brown anthers) male sterile plants are persistent until the seed ripens. Flowers are epigynous with five small sepals, five petals, five stamens & two carpels. Mature flower is 2 mm long. Nectars are present on ovary wall. Each carpel bears 2 ovule primordial, but lower one continues to grow. Embryo sac is monosporic (developing from chalazal macrospore) and 8 nucleate.

Fruit: Fruit is bilocular schizocarp, which dries and splits upon maturity to yield two mericarps (achenes) with one seed each. The mericarp are covered with spines.

Carrot has nine pairs of chromosomes with little variation in length. In that four are metacentric, four are submetacentric and one is satellite chromosome. There are 22 recognized species of *Daucus*. Most of these are with 11 or 10 chromosome pairs and only 2 other with 9 pairs. One interspecific hybrid between *D. carrot* and *D capillifolius* has been synthesized (Mc Collum, 1975).

Controlled Pollination: Flower size is small and seed yield per pollination is very less and hence very laborious. Anthers are removed from the early opening outer flowers in the outer whorl of umbellets until enough have been emasculated to ensure the needed supply of seed. Emasculation should be done before stigma become receptive. Unopened central florets in the emasculated umbellets and all late flowering umbellets are removed, leaving the female parent inflorescence with only emasculated flower. This umbel is then isolated under small cloth cage with a pollen bearing umbel from the selected male parent. Live house flies and pupae are introduced into the cage to ensure continuous pollen supply. Alternatively, without emasculation the anther or few umbels of selected parents are put in cloth cages and seeds are collected from both the parents and sown. Seedlings of F1s and parents can be distinguished based on vigour (roots and plant height), markers, etc.

Breeding History: Until 1960 nearly all cultivars were derived by selection in open pollinated material. Because of rapid loss in vigour due to selfing, little effort has been made to achieve uniformity by inbreeding. With discovery of CMS it was possible to develop F1 hybrids (1960 onwards). By 1983, 50 hybrid cultivars had been introduced in USA alone. State Breeding Program which uses to supply majority of parents to seed companies had started abandoning the carrot breeding program from 1970 to complete halt by 1980. This loss is not filled by major breeding efforts by seed companies because small seed companies which cannot back their breeding program have disappeared. Little competition and few hybrids came out with little demand.

Genetic Sources

Resistant to CMV	:	Virginia Savoy
Resistant to Motley virus	:	Sweet Crop, Top weight
Resistant to *Pythium* sp.	:	Nantes, Imperator
Resistant to Namatode (*Heterodera carotae*)	:	Vilmorium 66
Free from tip burn (physiological disorder)	:	Scarlet, Nantes.

Genetics

Root shape : Governed by three genes	:	D,N,P
1 Long or Desi : Long and tapering, D-N-P-	:	Eg. Pusa Desi
2 Cylindrical type: Cylindrical, dd, nn, P-	:	Eg. Nantes
3 Chantenay type: Obovate / cuneate (wedge shaped root) dd, N-P-	:	Eg. Chantenay
4 Round type: Round root, dd, N-, pp	:	Eg. Round variety

Genetics and Genetic Resources

Gene symbol	Character description	Gene source
A	α carotene synthesis	Kintoki cv
Ch	DM resistance	*D. carota* ssp. *dentatus*
Rs	Reducing sugar in root	——————-
'y'	Yellow xylem	——————-
O	Orange xylem	——————-
G	Green petiole	Tendersweet cv.

Root Colour

Digenic : P-Y- and two modified E_& I_

Deep purple:	iiPPYYEE	3 dominant genes
Purple :	iiPPYYee	2 dominant genes
Diffused purple :	iiPPyyee	1 dominant genes
Yellow :	IIppYYee	2 dominant genes
Red :	iiPPyyEE	2 dominant genes
Light red / orange:	iippyyee	three recessive genes

Root tip: Monogenic (S-), Blunt root is incompletely dominant to pointed root.

Root cracking: monogenic, susceptible dominant to resistance.

Male Sterility

Functional male sterility: Closed anther mutant reported.

Structural male sterility: Stamen less mutant reported.

It is controlled by plasma gene S and nuclear genes (a and B).

Action of a and B disappear when complementary factors C and D are present.

Duplicating gene action also reported.

Cytoplasmic Male Sterility

Two types of male sterility reported are

1. Petaloid type of male sterility: Characterized with poor seed yield. It is stable and widely used in hybrid programs. Stamen are replaced by 5 petals discovered by H.M. Munger in 1953 in 'Cornel Wild Carrot'
2. Brown anther type of male sterility: Characterized with good seed yield but not stable. Anthers degenerate and shrivel before anthesis. Source cv. Tendersweet. Not very stable (high seed yield) first report by Welch and Grimbal (1947) and Banga *et al.* (1964)

Eisa and Wallace (1969) reported petaloid type male sterility. Involving petaloid and brown anther types and three additional of brown anther type sources, Thompson (1962) studied inheritance of male sterility. At least 2 and probably 3 duplicate dominant maintainer genes and an epistatic restores operatively in the cytoplasm of both Tendersweet and Cornel Wild Carrot. Useful maintainer line should be free from restorer and homozygous dominant at one of the MS loci. Hansche and Gabelman (1963) reported digenic control of male sterility with either of two genes, dominant (MS-4) or recessive (ms-5), producing sterility interacting with cytoplasm from cv. Tendersweet.

Petaloid sterility in wild carrot cytoplasm is controlled by two dominant nuclear genes. In F_2 observed 15 fertile and 1 sterile plants.

Brown anther types: F_2 produce 15 sterile and 1 fertile: recessive controlled.

Various subsequent reports are available on male sterility stating varying genetics with varying sources.

Selection Techniques: Based on genetics, genetic resources and character profile

Uniformity: Market carrots in USA comprises nearly 80% are immature which are harvested and hence yield is not as important as uniformity.

High Pack-out: High Pack out include seed coating, precession plants, irrigation and fertilizer to provide uniform growing environment. Genetic uniformity, substantially contributes to the success of referred cultural practices. These are important even for processing carrots. If many hybrids are available for uniformity then competition starts for yield.

Appearance: Smooth and deep red orange colour in both xylem and phloem to eliminated green colour is preferred. Shape depends on consumer and grower preference.

Quality: Until recently eating quality has been neglected. Carrot is recommended for sources of provitamin A. Many consumers purchase carrot for their vitamin rather than taste.

Genetic Variation Reported for

Carotenoids: At least 3 genes (white to orange) and many more genes accounting for variation within orange category.

Orange carrot contain 40 ppm total carotenoids and of this 20 % α-carotene, 50 % β-carotene, 0 to 2 % lycopene and 0 to 10 % γ-carotene.

Besides carotenoids, fibre, texture, sugar, flavor, minerals and toxicants can decide the quality of the carrot. Dark orange carrots have high α-carotene i.e. up to 50%.

Colour: Selection should be carried out in uniform light under full day light or incandescent light. Under fluorescent light, roots appear much darker and a more desirable deep orange. Colour extending down to root tip is to be considered. Interior tissue colour also should be considered. For uniform interior colour, give transverse cut and there should not be green colour in cambium or upper portion of xylem. Recent hybrids have 120 to 150 ppm carotenoids. Open pollinated varieties have 80 to 100 ppm carotenoids. Red colour is due to lycopene. Visual selection is adequate for improving carotene content up to 120 ppm total carotenoids. Dark orange roots, the total carotenoids content may range from 130 to 200 ppm and for this laboratory analysis is must. Total carotenoids are determined analytically by comparing spectrophotometer transmission of 450 nm light through hexane extract of root tissue to the pure β-carotene standard solution. Hexane extract can be done by two methods.

1. Blend frozen raw carrot slices in a mixture of hexane and acetone. Then remove more polar acetone (along with water from carrot) in a separating funnel.
2. Lyophilize raw carrot slices and blend directly in hexane. This is rapid method.

Individual carotenes can be quantified by 1. Thin layer chromatography which is time consuming. 2. HPLC: High performance liquid chromatography: It is a rapid and accurate method.

Interior Quality

With transverse cut observe mature roots for 1. Cottony xylem, 2. Hollow heart, 3. Spongy phloem. Need to be observed under stress condition grown carrots using mature roots. Provide inducing conditions and follow selection.

Growth cracks: Cracks are more severe in processing cultivars when harvesting is delayed. It is also referred as harvest cracks or shatter cracks. It occurs when turgid root is subjected to mechanical abuse at harvest.

Sugar and Flavour: Organoleptic test can be used for selection. Reject the roots of bitter or harsh flavour. The selector must be able to detect flavour difference.

Total Soluble Solids: Put root juice to refractometer or take 5 g frozen sample seal in polyethylene held at -10°C and is transferred and macerated within sealed bag. A needle puncture will release the drops of juice needed for refactrometer reading. TSS, Specific Gravity, and dry matter are inter related. Sugar quality,

TSS and percentage dry weight exhibit quantitative inheritance pattern (Scheerens and Hosfield, 1976).

Total sugar and TSS or per cent dry weight have rp = 0.75 to 0.95

TSS and percent dry weight are highly correlated with rp = 0.85 to 0.95

TSS accounts for 90% of dry weight, where as total sugar accounts for only 60 to 75 % of soluble solids and 40 to 60% of dry weight.

Average raw carrot composition is 88% water, 6 to 8% sugar, 1 to 2 % fiber, 0.7 to 1.2 % protein, 1% ash and 0 to 0.3% fat and it is starch free. Sugar can be measured calorimetrically. But HPLC gives rapid sugar analysis. Increased sugar content improves carrot flavour. Volatile terpenoids and other potent compounds other than sugars have greater effects on carrot flavour. Carrot roots must be taste tested to achieve the correct balance between volatile terpenoids and sugar levels. Lack of consistent flavour improvement is observed by selecting for increased TSS. Dipping slices in 30% fructose could not enhance flavour and hence terpenoids have decisive role in overall flavour. Total sugar is sum of glucose, fructose and sucrose accounting to 98 to 100% of free sugar in carrot. The reducing sugar 'Rs' locus controls the ratio of reducing sugar (glucose and fructose) to sucrose (non- reducing) independent of total sugar (Freeman and Simon, 1983). Dominance is for high reducing sugar and low sucrose. The quality of deep fried carrot chips may be damaged by presence of high reducing sugar (Rs). Otherwise, high reducing sugar may be desirable for improving flavour of raw carrots. Since average sweetness of glucose + fructose is 20% greater than that of sucrose.

Terpeonoids: Selection for reduced levels of volatile terpenoids is of prime concern to improve flavour of raw carrot. Low volatile terpenoids provide blend but high level will not be edible. Terpenoid quantity and associated harsh flavour exhibit greater genetic variation. Volatile terpenoid vary from 5 to 200 ppm with little effect of environment. Volatile terpenoids at 20 to 200 ppm concentration are desirable. High level leads to harsh flavour and low level leads to lack typical carrot flavour. Inheritance of harshness is multigenic, early selection in segregating generation is desirable. With laboratory assessment for specific type of terpenoid one can establish elite population for flavours.

Isocoumarin contributes to bitter flavour and demonstrate quantitative genetic variation. It is the consequence of extended cold storage, particularly in presence of ethylene. Hence, it is not the prime concern in breeding. Role of isocoumarin in bitter flavour is not fully known.

Non-bolting: Expose segregating population to induction environments and imposing continuously selection pressure for a high vernalization requirement. Non bolting is required for cultivars for temperate region and for seed production it is not desirable and hence reasonable amount of bolting is desirable. Eight to ten weeks at temperature below 5^0 C are required for complete vernalization of steclking harvested in Michigan in late August and intended for seed production in the green house.

Inbred vigor: Very high inbreeding depression noticed in early germination itself. Many inbreds failed to survive or were discarded when they appeared to be too weak to reproduce. To increase seed yield in hybrids three way crosses attempted. With continuous efforts one can achieve tolerance to inbreeding and that inbred lines with enough vigour for direct use in single cross hybrids need to be developed.

Objectives

- High yield and earliness: High root weight
- Uniformity, Non bolting
- Heat tolerant lines
- Rich in carotene and good colour, shape and texture and smoothness with high sugar
- Free from disorders/defects like root branching, pithiness, cracking and tip burn
- Breeding varieties for nontraditional areas like warm humid tropics.
- Resistant to diseases: leaf spot, motley virus, CMV and nematodes.
- For improved Flavour, texture and nutritive value (modern objective).

Breeding Methods: Andromonoecy, protandry of perfect flower, male sterility and self incompatibility prevailing in carrot favours cross pollination and accordingly methods can be adopted.

Introduction: Imperator, Danvers, Nantes and Perfection were introduced to India.

Mass selection: Followed for improvement of root length and getting high yield.

Bulk population methods: Pusa Kesar is the selection from cross between Nantes x Local Asiatic by following bulk population method.

Mutation breeding: In mutation breeding 0.025 – 0.06% NEU (N-nitroso-N-ethyl urea) and 0.012% NMU (N-nitroso- N-methyl urea) were successfully used to develop male sterile lines.

Polyploidy breeding: Tetraploid (2n = 36) and octoploid (2n = 72) have been developed but has limited utility.

Heterosis breeding: Heterosis is reported for earliness, tuber length, tuber yield, carotene content, top weight, core diameter and root diameter. Male sterility and self incompatibility are used to exploit heterosis. From private seed companies commercial hybrids are available.

Interspecific hybridization: Resorted to generate useful variability

	D. c. ssp sativum	*D. gingidium*	*D. capillifolium*
D. carota ssp. sativum	SI	CC	CC
D. gingidium	CC	SI / SC	NK
D. capillifolium	CC	NK	SI / SC

SI – Self incompatible, SC – Self compatible, CC – Cross compatible NK – Not known

Breeding plan: F_1 and maintainer lines with life cycle require1 year duration.

Year	Season	Cycle	Activity
1	a	V	Grow and select roots from a diverse collection
2	a	R	Self pollination and cross to a cytosterile tester as many as possible
	b	V	Select 5-10 roots of the best S_1 lines with their companion F_1 test crosses
3	a	R	Pair 5-10 individual roots from each selected line with its F_1 under small cage isolators to produce S_2 and BC_1 seed
	b	V	BC_1 roots selected to resemble their S_2. Store and vernalise. 40 to 50 BC_1 roots to use in progeny tests for maintainer genotypes
	a	R	Produce S_3 and BC_2 seed
	b	R	Classify 40 to 50 BC_1 plants as ms or mf in progeny test to identify maintainer lines
5	a	R	Mass 15 or more S3 roots of maintainer (M) line and 5-10 phenotypically similar selections from the companion BC2 (S) lines
	a	R	Produce experimental F1 hybrid combinations.
	b	V	Multilocation testing of F1S
	b	V	Produce roots from BC3 and S3 mass selection to produce 20-40 roots of candidate line indentified in the hybrid trials
6	a	R	Isolate under large cages the roots to produce bc4 sterile (s) and s3m2 maintainer companion lines of the best inbreeds
			Continue till you get iso-genic maintainers and consistent perfmonce of F1 hybrids observed.

V: vegetative cycle R: reproductive cycle

Disease Resistance

1. Alternaria leaf blight: Caused by *Alternaria dauci*

   USA Sources: 1. Kokubu Japanese cvs. (PI 261)
   2. Imperial long scarlet
   3. Sanm Nai PI 226043 (Japanese cv.)

2. Cercospora leaf blight: Caused by *Cercospora carotae* and prevalent in USA

3. Aster yellows: MLO disease transmitted by six spotted leaf hopper. Prevalent in North USA

4. Carrot motley dwarf: Prevalent in UK : Induced by two viruses viz., Carrot mottle virus and carrot red leaf and transmitted by willow aphid.

Salient Breeding Achievements

Pusa Meghali: A carrot variety suitable for halwa preparation.

Arka Suraj (Selection 9): Pedigree selection from cross between Nantes X IHR 253, deep orange roots with self colour core, smooth root surface, conical shape, root length 15-18 cm, root diameter 3-4 cm, TSS 8-10% , carotene content 11.27 mg/100g, tolerant to Powdery mildew. It flowers and set seeds under tropical conditions. It yields 24 to 25 t/ha.

Chaman: This is temperate variety, developed at SKUAT, Srinagar. Plants are with dark green and semierect foliage. Roots are long, cylindrical semi- blunt, tolerant to cracking and forcing with orange, sweet, fine textured flesh, gives yield of 25-27 t/ha.

Nantes: This is a temperate variety, developed at IARI Regional Station, Katrain. Plants are small with green leaves. Roots are slim, small, orange, tender, sweet, well shaped with small self coloured core, cylindrical and abruptly ending in a small thin tail. Flesh is orange scarlet. Suitable for sowing from mid October to early December; gives average yield of 15-20 t/ha in 90-110 days of crop duration.

Pusa Kesar: This variety is adapted to sub-tropical climates and developed by IARI, New Delhi. Roots are deep red, tapered, small and self-coloured core, rich in beta carotene, gives an average yield of 25 t/ha in 100-120 days of crop duration. This is popular in J&K, HP, Uttaranchal, Punjab, UP, Bihar, Jharkhand, MP and Maharashtra.

Pusa Yamdagini: This is a temperate variety developed at IARI, Katrain. Roots are 15 to 16 cm long, orange with self coloured core, slightly tapered and semi stumpy with medium tops, rich in carotene, gives an average yield of 9 to 10 t/ha in 90 to 100 days of crop duration. This is popular in J&K, HP and Uttaranchal.

Pusa Rudhira : Roots mature at 75-90 days after sowing with yield of 30 t/ha. Roots long red with self-core, juicy, sweet, suitable for salad, juice extraction, candy making, puddle, pickle and cooking. Roots high in health nutrient content (/100 g), total carotenoid (7.6 mg), a-carotene (4.92mg), lycopene (6.7 mg), zinc (10.7 pg/g) and iron (230 pg/g).

Pusa Asita: It has black roots (anthocyanins) which are juicy and sweet. It is suitable for making health drink kanji, salad, puddle and color extraction. The foliage dark green, root attain 21.54 cm in 90 to 110 days. It yields 27 t/ha. It is rich source of anthocyanins (520 mg), total carotenoids (0.57 mg), lycopene (0.52 mg), zinc (8.45 pg/g), iron (260.60 pg/g) and calcium (368.67 pg/g).

22

Radish

It is a popular root crop having wide range of adaptability, grown in Japan, Koria, China, India and other European countries.

Origin and Distribution: Radish was an important food in Egypt during 2000 BC. Fossils study reveals that, it was a food as early as 2700 BC. It spread to China during 500 BC and to Japan during AD 700. Variability of crop decreased from Europe to China and from there to Japan. Recent study indicates that East of Mediterranean, most probably China was the centre of origin of the crop. This means Europe and Asia are the probable places of origin.

It is true diploid with 2n = 18

Genetics

Character	Number of genes	Type of gene action
Leaf lobbing	Monogenic	Lobbed leaf dominant to simple
Leaf hairiness	Monogenic	Hairy leaf dominant to glabrous
Root colour	3 genes and complimentary gene action	Red : R_2r_2 /R_3r_3 /cc White : r_2r_2/r_3r_3 /Cc Purple : r_2r_2/ R_3r_3/cc R_2r_2/r_3r_3 /Cc R_2R_2/R_3R_3/CC : lethal and can not survive
Leaf colour	Monogenic (YgYg)	Dark green leaf colour dominant to yellow green

Genes	Characters
a	Plant entirely lacking anthocyanin. Petals white, seeds yellow
AC1	White rust, Race 1 resistant
ms1	Male sterility, absence of pollen
Pi	Pink pigmentation in hypocotyls, stems, and other plant parts
Pu	Purple pigmentation, strong purple in siliqua
RS	*Rhizoctania solani* resistance
gf1, gf2	Green streak or spots or flecks on leaves

Objectives of Breeding

- High yield and earliness
- Varieties adopted to different growing conditions
- Improved qualities-shape and size of root, free from branching, cracking, etc.
- Resistant to pest and diseases: Leaf eating caterpillar (*Erioschia brassica*), white rust, wilt, black rot, downy mildew, leaf spot and mosaic diseases.

Breeding Methods: It is a cross pollination crop and mass selection is well suited.

Mass Selection: Arka Nishant developed through mass selection by using stock IIHR 72.

Pedigree Method: Pusa Himani developed through pedigree method involving the cross Black Radish x Japanese White.

Pusa Rashmi developed through pedigree method involving the cross Green Top x Desi

Polyploidy Breeding : Polyploids with 2n = 36 produced and there were no distinct advantages observed. Two polyploid varieties have been developed and these yielded more than diploids. These are Sofia Delicious (2n = 36) and Semi long Red Giant (2n = 36)

Interspecific Hybridization: *Rhaphanus indicus, R. sativus, R. sensustricta* and *R. raphanistroides* are completely cross compatible. This was done to develop heat tolerant lines and resistance to diseases and pests.

Intergeneric Hybridization: *R. sativus* crosses with *Brassica oleraceae* when radish is used as female parent. This was done to transfer male sterility into cabbage and cauliflower.

Heterosis Breeding: Heterosis is observed for yield, root weight and Vit C content. Some hybrids developed which are of single cross, double cross and three way crosses.

Breeding Achievements

Asmer Tip Top, Sparkler: Resistant to insect *Erioischia brassicae* (Leaf eating caterpillar)

Scarlet Globe, Rapid Red: Suited for spring season

Japanese White: This variety has been developed by IARI Regional Station, Katrain. Roots are thick, 20-30 cm long, plumpy, cylindrical, mild pungent, smooth and white with green shoulder. Flesh snow-white, crispy, solid, mild pungent.

Suitable for sowing between October to December in plains and during September in hills and gives yield of 15 to 30 t/ha in 60-65 days of cropping duration. This is popular among the farmers of J&K, HP, Uttaranchal, Punjab, UP, Bihar, Jharkhand, MP and Maharashtra. Seeds could be obtained from the branches of NSC.

Pusa Himani: This variety has been developed by IARI Regional Station, Katrain. Roots are white, medium thick, 30 to 35 cm long, tapering with white green shoulder. Flesh is white, crisp and sweet flavoured with mild pungency; suitable for sowing throughout the year except winter months in hills. It gives an average yield of 32.0 t/ha in 60 to 6 5 days of crop duration. This is popular among the farmers of J&K, HP, Uttaranchal and West Bengal. Seeds of this variety could be obtained from the branches of NSC.

Punjab Safed: It is resistant to leaf spot

Black Radish and **Pusa Chetki** are suited for summer.

Chinese Rose White is resistant to Downy Mildew.

Nerina : It is resistant to Mosaic virus.

Summer Wonder: It is resistant to Black rot.

Palam Hriday : Radish variety developed at HPAU, Palampur and is pink fleshed.

Pusa Chetki (Exotic variety from Denmark): This has been developed at IARI, New Delhi. Roots are white, smooth, soft in texture, less pungent, 12 to 20cm long and stumpy. Leaves are with entire slightly lobed, dark green and upright. It gives an average yield of 25 t/ha in 45 to 50 days of crop duration. This is popular among the farmers of J&K, HP, Uttaranchal, West Bengal, Assam, Punjab, UP, Bihar, Jharkhand, MP and Maharashtra.

Kashi Hans: This has been developed through selection. It is suitable for September to February planting and harvesting can be done after 40 to 45 days of sowing. It can stand in the field up to 10 to 15 days after commercial maturity. Leaves are soft and smooth like spinach. Roots are straight, tapering, 30 to 35 cm long, 3.5 to 4.2 cm diameter. It yields 430 to 450 q/ha. This has been identified by UP State Horticultural Seed Sub Committee and notified during the XII meeting of Central Sub Committee on Crop Standard Notification and Release of Varieties for Horticultural Crops for U.P., Punjab, Jharkhand.

Kashi Sweta (IIVR-1): This variety has been developed by IIVR, Varanasi. It has been derived through selection from a Chetki population. This variety is suitable for early harvesting (30-35 days after sowing). Roots are 25 to 30 cm long, 3.3 to 4.0 cm in diameter, straight, tapering with pointed tip, gives an average yield of 45 t/ha.

White Icicle: This variety has been developed at IARI, Regional Station Katrain. Plant top is medium short, root pure white, thin, tender, solid, straight and tapered. Flesh is icy white, crispy, mildly pungent and flavoured. It is popular among the farmers of HP and Uttaranchal.

23

Beet Root

Beta vulgaris 2n = 2x=18 is an important vegetable crop in Eastern and Central Europe and it belongs to family *Chenopodiaceae*. It is of lesser importance in Western Europe and USA where it is known as garden beet. The swollen roots are consumed as vegetable and salad. The other types which are non-horticultural forms are known as sugar beet, mangold and fodder beet.

Origin: Leaf beet is the earliest domesticated form. It is believed to be originated from *Beta maritima* known as sea beet which is indigenous to Southern Europe.

Botany: Biennial producing enlarged hypocotyl (roots) and rossette of leaves in first year and flowers and seeds in second year.

Genetics

Root colour: At least two genes involved. Red is dominant to yellow or white. Intensity of red colour is influenced by minor genes. Further, there are 5 alleles at R locus: R, Rt, r, Rp and Rh and 3 alleles at Y locus: Y, Yr and y control the colour.

R_Y_ : Red roots, hypocotyls and petioles

rr Y – : Yellow roots, petioles and hypocotyl

R_yy : White roots with red hypocotyls

rryy : White roots and yellow hypocotyls

Male sterility: Gene 'X' in S cytoplasm gives fertility. Fertility is dominant. Second gene 'Z' causes partial male fertility. This is dominant to male sterility and independent and hypostatic to 'X'.

Objectives

- High root yield
- Dark red, uniformly coloured roots
- Uniform root shape
- Absence of internal white rings in roots

- Slow bolting
- Monogerm seed
- Resistance to Downey Mildew (*Pernospna farinoss*) and Powdery Mildew (*Erysiphe communis*).

Breeding Methods: It is cross pollinated crop has considerable inbreeding depression and suited methods of breeding are:

1. Mass selection and progeny testing
2. F1 hybrids with cytoplasmic male sterility
3. Mutation breeding: Dimethyl sulphate (DMS), N-methyl-N-niroso urea (MNO), N-ethyl N-nitroso urea (ENU), r-rays and UV rays used to induce variation.

Achievements

"**Avonearly**" bred by selection from Detriot type.

Crimson Globe and **Detriot Dark Red** were introduced and are popular in India.

24

Turnip

Turnip is the root vegetable cultivated in cooler season or temperate regions of the globe. *Brassica rapa* L 2n = 2x = 20 is formerly known as *B. campastris* subsp *rapifera*. Swede (*B. napus* var. *napobrassica)* is directly related to turnip.

Origin: It is originated in cooler part of Europe, presumably from biennial oilseed forms. It was known to Greeks and Romans at beginning of Christian era. It was introduced to Britain from France by Romans and into North America by the early European settlers in the 17th century.

Genetics

Sporophytic system of Self Incompatibility operates.

Flesh colour: White is dominant to yellow

Skin colour: Two independent loci, both dominant genes give purple colours

Bolting: Two major additive genes operate.

Club root: Three independent dominant genes operate.

Powdery Mildew: Qualitative inheritance: recessive allele has partial resistance.

Turnip mosaic virus resistance: Single dominant gene controls resistance.

Hybridization: Self Incompatibility operates where one has to enclose heads of two compatible plants and use blow flies.

Selfing: NaCl and CO_2 can be used to overcome self incompatibility

Objectives of Breeding

- Earliness
- Root colour depending on preference white and purple types can be developed
- Stump root, non branching roots
- Slow bolting
- 8 to 9 per cent dry matter in roots

- Resistance to club root, powdery mildew, turnip mosaic virus, white rust, phyllody, cabbage root fly and turnip root fly.

Methods of Breeding: It is a cross pollinated crop and breeding methods suited for cross pollinated crops can be followed for improvement.

Mass Selection: To develop open pollinated varieties by massing of selected root plants.

Heterosis breeding: Successful in white turnips. SI system used.

Varieties Developed

Biennial or temperate or European types

Annual or Asiatic types

Temperate Types

1. Pusa Chandrima : At Katrain derived from Japanese White x Snow Ball
2. Pusa Swarnima: It is derived from cross between Japanese White and Golden Ball at Katrain. It is less susceptible to turnip malformation and MLO disease.
3. Purple Top White Globe: It has round root, purple top and white bottom
4. Golden Ball: It is susceptible to turnip malformation
5. Snow Ball : It has globe shaped roots which are white in colour.

Asiatic Types

1. Pusa Kanchan: It was derived from cross between Local Red Round (Tropical type) x Golden Ball Red (temperate type)25 to 30 t/ha
2. Pusa Sweti: It has white roots and yields 20 to 30 t/ha.

25

Sweet Potato

Sweet potato *Ipomea batatas*(L). Lam. is an asexually propagated vegetable. It is grown in most of the tropical and subtropical region of the earth. In USA production extended to temperate regions. Vines as well as roots are eaten or fed to live stock. It is the source of industrial starch and potential source of the alcohol. Among the major tuber crops cultivated in India sweet potato rank third after potato and cassava in area and production.

Origin: Available evidence suggests South Mexico through Central America and Northern South America as the probable center of origin. It is perennial but handled as annual. Its hardiness and drought tolerance are well recognized.

Family: *Convolvulaceae,* Chromosome number: 2n=90

It is the only known natural hexaploid morning glory. Most of the wild species are diploid (2n=30). Hybridization is restricted with in species because of its hexaploidy nature. Extensive variability with in species is observed. Tuber is the fleshy subterranean stem or shoot but in sweet potato tuber is a root, and hence it cannot be a tuber (common error). It is a perennial, and the storage root is capable of continued enlargement and does not mature in the sense of reaching some final size or stage of development. From the stand point of function, the root system consists of absorbing roots and fleshy storage roots.

Sweet potato might have originated through following evolutionary events

a) Occurrence of *Ipomea trifida* (6x) the only wild species having 2x, 3x, 4x or even 6x by duplication of *I. leucantha* (2x) or close relative.

b) Domestication of wild type characteristic of *I. trifida* (6x)

I. leucantha (2x) x *I. littoralis* (4x)

Diploid(DD) ↓ Tetrploid (BBBB), wild type

hexaploid (BBBBDD)

Domestication of *I. batatas*(6x) led to evolution of cultivated types.

Tuberous root forming diploid and tetraploid relatives have still not been found.

Floral Biology: Flowering occurs in axillary inflorescence of 1 to 22 buds. Colour of flower range from white to lavender. Tube depth varies from 28 to 63mm. Five petals are fused and have stamens attached at their bases with anther that are normally white, but may be light or dark lavender. Filaments vary in length from 5 to 21 mm, and this affects position of anther in relation to stigma. Style 8 to 29mm long, there are 2 ovaries in the pistil each containing 2 ovules, sepals are leaf like and persistent and may be glabrous or pubescent. Base of corolla consists with yellow glands that contain nectar which attracts insects. Capsule contain 1 to 4 seeds and may be glabrous or pubescent, 100 seed weigh is 1.3 to 3.0 g. Seeds are hard and may retain viability for 20 years or more. Germination is irregular and needs scarification. Soak seeds in concentrated H_2SO_4 for 20 to 60 min and washed and neutralized with solution containing bicarbonate of soda and rinsed in clean water. Wide genetic differences in flowering incidence as well as strong environment influence. You can select for flowering character because no undesirable trait associated with good flowering.

Germplasm Collection: Following are the centers where potato germplasm is maintained.

1. International potato Center (CIP), Peru donated by different institutes 1895 acc.
2. CTCRI, Trivandrum, India collected 860 acc.
3. CIP, Peru (Lima)- 3993 sweet poatato. In this wild batatas 301, wild other genera 406, wild undetermined -412

Besides these centers GAAS (China), USDA (USA), Japan, Indonesia, Papua New Guinea are the countries where sweet potato improvement is carried with available germ plasma collection.

Related species grouping

A group(triloba group)	X groupBridging group	B group(*I. batatas* group)
I. triloba (AA) *I. lacunose* (AA) *I. ramoni* (AA) *I. tricocarpa* (AA)	*I.tiliacea*(4x) *I.gracilis*	*I.leucantha (BB)* *I.littoralis* (BBBB) *I.trifida* (BBBBBB) *I.batatas* (BBBBBB)

1. 'A' group: All the species in the group are self-compatible and cross-compatible with each other and compatible with species of 'X' group but incompatible with species of 'B' group.
2. 'X' group: Species in the group are self incompatible but cross compatible with each other and are compatible with species of both 'A' and 'B' group.

3. 'B' group: All the species in the group are self incompatible but cross compatible with each other and incompatible with species of 'A' group.

Major Objectives

- Yield
- Earliness
- Tolerance to sweet potato weevil
- Quality (increased carotene)

The improvement programmes taken on short term and long term projects. To achieve the objectives various breeding methods have been adopted in the crop improvement

Breeding Methods: Poly cross and mass selection technique appears to be most efficient way to breed sweet potato because technique of controlled crossing will not be repeated (here per pollination only 4 seeds). Insect problem for flowers can be overcome by spraying in the evening hours. Insect pollination is chiefly by insects of hymenoptera besides honey bees and bumble bees.

Mass Selection: A comprehensive sweet potato breeding programme addresses both long and short term goals should contains one or more mass selection population to provide new types.

Long term goals	Short term goals
1. Parents	1. Cultivars
2. Average line values	2. Individual plant values
3. Large no. of plants	3. Small no. of plants
4. Less precise evaluation	4. Precise evaluation
5. Gradual and continuous improvement	5. Quick improvement
6. Fundamental research	6. Adaptive research
7. Transmission	7. Expression
8. Mass selection procedure	8. Back cross procedure

Original plant material come from widely divergent sources and sources are introgressed as available

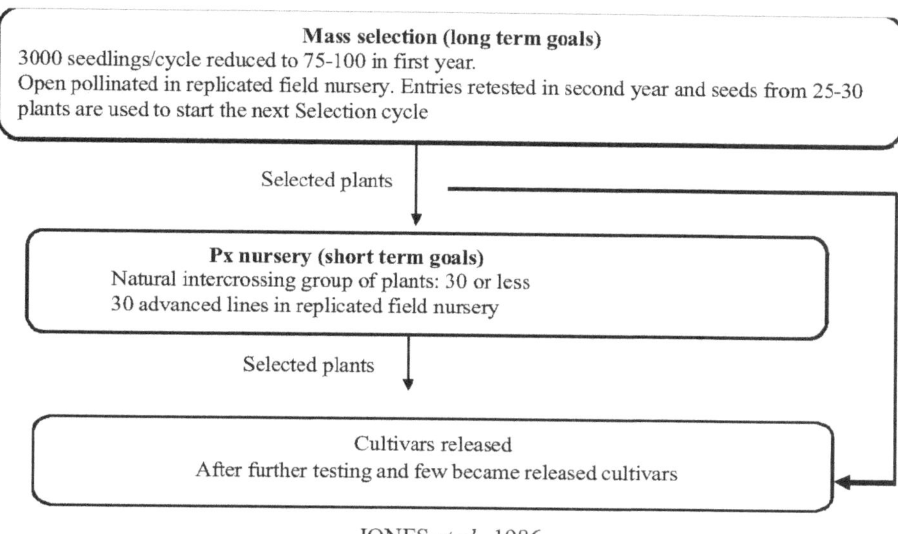

JONES *et al.*, 1986

Clonal Selection

CO-1: It is the clonal selection from indigenous collection made at TNAU, it requires 120 days for crop maturity after planting, it gives 18 to 22 t/ha yield, skin colour is light pink, it has white flesh and is field tolerant to sweet potato weevil.

VL Sakarkand-6: Exotic collection B219 from USA, released for general cultivation in hilly regions of UP, elongated tubers with purple skin and light yellow flesh, tuber length is 15 to 18cm with 12 to 14cm girth. It yields 18 to 22 t/ha.

Open Pollination and Selection

1. Sree Nandini (75-0P-217): It has fusiform tubers, light cream skin and white flesh. It yields 20 to 25 t/ha in 100 to 105 days. It is highly tolerant to sweet potato weevil.
2. Sree Vardhini (76-0P-219): It yields 20 to 25t/ha in 100 to 105 days. Roots are rich in â carotene (1200 IU/ 100 g edible part). It is highly tolerant to sweet potato weevil and nematode.
3. CO-2: Variety matures at 120 days and yields 20 to 25 t/ha
4. CO-3: Variety matures at 110 to 120days after planting. It has light red skin colour with dark orange flesh, with high carotene content.
5. Samrat: It is early harvest type and gives 18 to 24 t/ha yield. It is moderately resistant to sweet potato weevil.

Hybridization and Selection

H-41 : It is the cross between Norin (Acc no. 1120) from Japan and Bhadrakali Chola (Acc no. 3) a local collection. It matures in 120 days. It has white skin with white flesh and yields 20 to 25 t/ha. It is highly tolerant to sweet potato weevil.

H-42: It is the cross between Valla Damph (acc no. 1871) a local collection and Trimph (acc no.1130) from America. It has pink skin with yellow white flesh, tolerant to sweet potato weevil.

H-268: It is double cross hybrid involving 4 cultivars. It is red skinned with pink rind and light yellow flesh. It yields 18 to 20 t/ha in 120 days. It is tolerant to sweet potato weevil.

Rajendra Sakarkand-5: It is derived from cross 4 x M-5. It yields 20 to 25 t/ha in 120 days. It is tolerant to sweet potato weevil.

Breeding Achievements: There are many varieties, which are differentiated mainly by the shape, size, colour of tuber, flower colour and leaves. However there are two important groups of varieties from the market point of view, based on outer colour (rind) of tuber. They are usually grouped either as white or red type. White type is supposed to be less sweet and more fibrous whereas red types are shorter in duration and has a better quality flesh. Varietal types are based on the characteristics of the cooked flesh, namely a) Varieties with firm, dry, meaty flesh when cooked and b) Varieties with soft, moist, sugary flesh when cooked.

Konkan Ashwini: It is suitable for both kharif and Rabi season. It developed by using the genotype palghar-1 collected from Varor village in Thane dist. of Maharastra. It has long elliptical tubers, deep purple and cream coloured flesh with good consumer preference. It yields 12.38t /ha in kharif and 19.22t/ha in irrigated conditions.

V-8 (FB-4004): Crop duration is of 4 months. The vines are long and trailing, with a purple shade at the internodes. The stem is round smooth and hairy. Tubers are small in size, spindle shaped and white skinned. The flesh has a creamy colour. It is non fibrous and agreeable to taste.

V-12 (TST white or Tie Shen Tun): It matures in three and a half to four months duration. The vine has a semi erect habit. The tubers are big in sized and elongated and have white skin. The flesh is white, it is sweet and is non fibrous.

Pusa Suffaid (IARI): This is a selection of a white variety originally introduced from Taiwan. It is a high yielding table variety with wide adoptability. It is more popular in Delhi, Punjab, Rajasthan, UP, Bihar, Maharashtra and Tamil Nadu. The tubers are somewhat elongated, white skinned and white fleshed. The boiled flesh is creamy white, moist, sweet and very palatable. It matures in 105 days and gives 26 to 30 t/ha. It is less affected by weevil.

Pusa Sunehri (IARI): It is a selection from material obtained from the USA. The tubers are elongated, light brown skinned and yellow fleshed. The boiled flesh is attractive orange yellow, moist and sweet. The tubers contain 24.8 ppm of carotene.

Pusa Lal (IARI): This is a selection from a Japanese variety (Norin). It has purplish red skin and white flesh and produces a high proportion of marketable tubers. It is popular in Delhi, Punjab, Rajasthan, UP, Maharashtra.

Sree Nandini (75-0P-217) (CTCRI): It is a selection from the open pollinated progeny. The tubers are fusiform with light cream skin and white flesh. It yields 20 to 25 t/ha in 100 to 105 days after planting. Roots cook well and are tasty. It is highly tolerant to sweet potato weevil.

Sree Vardhini (76-0P-219) (CTCRI): Sree Vardhini was selected from the progeny of open pollinated seeds. Tubers are rounded with pink skin and light orange flesh. It yields 20 to 25 t/ha in 100 to 105 days. Tubers cook well and have good acceptable taste. Carotene content of tubers is about 1200 IU. It is highly tolerant to sweet potato weevil and nematode.

CO-1 (TNAU): It is a selection from indigenous collection made at TNAU, 1976. It yields 18 to 22 t/ha in 120 days. It skin colour is light pink with white flesh. It is tolerant to sweet potato weevil.

CO-2 (TNAU): It is a seedling selection from open pollinated progeny. It yields 20 to 25 t/ha in 125 days.

CO-3 (TNAU): Duration of this variety is 110 to 120 days. Its skin colour is light red with dark orange coloured flesh indicating high carotene content.

V-2 (PAU): It has broad leaves which are dark green in colour. Stem is long, thick and hairy. Tubers are large in size, white in colour and non fibrous. It yields 15 t/ha.

V-6 (PAU): It has purple colour leaf baser. The tuber is large, red in colour and non fibrous. It yields 10 t/ha.

V-8 (PAU): Leaves are broad, entire and dark green with purple tinge. Base of leaves is havaing purple stain. The stem is thick, long and purple in colour. Tubers are long, thick, white in colour with red hue and slightly fibrous in cooking. It yields 6 to-11.8 t/ha.

Cross 4 (RAU, Bihar): It is released in mid sixties (1965). Matures in 105 days and 20 to 30 t/ha. It is highly susceptible to weevil infestation resulting in 30 to 40 per cent reduction in yield.

Rejendra Sakarkand-5 (RAU, Bihar): A hybrid developed from the cross 4 involving cross 4 x M-5. Tubers are cylindrical and per plant can bear 4 to 6 tubers. Each tuber can weighing around 500 to 600 g. It gives 20 to 25 t/ha yield in 105-120 days. The colour of tuber and flesh is white. The tubers cook well with accepted taste. It is suitable for September (autumn) planting under North Bihar conditions. It is resistant to *Fusarium* wilt and *Cercospora* leaf spot diseases and tolerant to sweet potato weevil.

Kalmegh (RAU, Bihar): It is a selection with extra early bulking. It yields 26 to 32 t/ha in 90 days. It could be easily fitted in crop rotation in a year with maize, sweet potato, wheat and moong (Green gram).

Samrat (APAU): It is the selection from the open pollinated seedling progeny. Tubers are *fusiform* with light pink skin and white flesh. It is early maturing. It yields 18 to 24 t/ha. It is moderately resistant to sweet potato weevil.

VL Sakarkand-6 : Tubers are elongated with purple skin and light yellow flesh. Tuber length is 15 to 18 cm with 12 to 14 cm girth. It yields 18 to 22 t/ha. Tubers cook well and are tasty.

H-41 (CTCRI): It is a hybrid between a Japanese variety Norin and Local variety Bhadrakalichola. Tuber size is 16 to 20 cm in length and 4 to 5.5 cm in diameter. Tubers are *fusiform* with white skin and flesh. It yields 20 to 37 t/ha in 120 days. Tuber boils easily and has sweet taste with less fibre. It is highly tolerant to sweet potato weevil.

H-42 (CTCRI): It is a cross between an indigenous variety Vella Damph and American variety Triumph. Tubers are 14 to 20 cm long and 4.6 mm diameter and are *fusiform* with pink colour skin and yellowish white flesh. The cooked tubers are sweet in taste and free from fibre, It yields 20 to 37 t/ha in 120 days. It is highly tolerant to sweet petals weevil.

Vikram (CI 853-4): It is a selection made at UAS, Dharwad from the plant introduction from AVRDC Taiwan. Tubers are pink skinned, flesh is creamy, matures in 85-95 days (short duration) and escapes sweet potato weevil. It yields 35 t/ha. It is tolerant to water stress condition.

H-268: It is a double cross hybrid. It yields 25 to 28 t/ha in 120 days. Tubers are *fusiform*, long, red coloured with pink red and light yellow flesh. It is highly tolerant to weevil, popular in Maharashtra.

S-30: Duration is 90-110 days. Tubers are round and light brown. It yields 22 to 30 t/ha and is susceptible to weevil.

V-35 : Duration is 105-120 days. Tubers are round, light brown and less affected by weevil and it yields 26.5-30 t/ha.

8-635: Duration is 120-140 days. Tubers are long, red, less affected by weevil and yields 20 to 25 t/ha.

H-620: Duration is 120-135 days. Tubers are long, red, less affected by weevil and yields 22 to 25 t/ha.

R-S-5: Duration is 105-120 days. Tubers are *fusiform,* white and it yields 27.5-32.5 t/ha.

Exotic Cultivars: In the USA, sweet potatoes are classified as food types and feed types. Usually the feed type produces higher tuber yield and high starch. Food types are classified into two types based on the characteristics of cooked sweet potato.

a) Firm fleshed (soft fleshed greater commercial importance than the firm - fleshed type). Firm fleshed cultivars are Big stem Jersey, Yellow Jersey - Old cultivars, whereas Maryland Golden, Orlis and Jersey Orange.

b) Soft fleshed cultivars are Puerto Rico, Unit-I and Cleitt Buneh Puerto Rico, Nancy Hall, Ranger, Early Port, Goldrush, Heart-a-Gold, All gold, Australian Canner and Triumph.

Table: Varietal description

Cultivars	Duration (Days)	Yield (t/ha)	Colour Skin- Flesh	Salient feature
Bhuban sankar	110-120	15-20	red-white,	Good cooking quality
CO-1	120-135	20-25	Light pink-white	
CO-2	110-115	20-25	Flesh white,	Cooks well
CO-3	105-115	20-28	red –orange,	rich in carotene
Gouri	110-120	18-25	red- orange	rich in carotene
H41	120-130	20-30	red- white	
H42	120-130	20-25	Pink tubers	
Kiran	110-120	20-25	red-orange	Rich in carotene
Rajendra Sakarkand-5	105-120	24-30	white-white	Tolerant to weevil
Rajendra Sakarkand-35	105-120	25-30	Brown –white	
Rajendra Sakarkand-43	110-120	20-23	Brown-white	Mod. Res. to *Cercospora* leaf spot, Tolerant to weevil

26

Cassava

Cassava (Syn: Tapioca) *Manihot esculenta* (Crantz) (2n = 2x = 36) (Hindi : *Mravuli)* is perennial shrub cultivated in tropics for its starchy tuberous roots. It is one of important tuber crops cultivated in tropics because of its high carbohydrate yield per unit land and labour and its adaptation to poor soil and water stress and its tolerance to dominant pest and diseases. Cassava contains cyanide, the toxic principle which limits food and feed value. In cassava the glycoside linamarin and lotaustralin are converted to hydrocyanic acid, a potent toxin, when come in contact with linamerase, an enzyme that is released from rupture of cells. Tuber peel contain significantly higher amount of glycosides than the pulp. It can be removed by cooking in water for about 5 minutes.

Uses

- Human food (60%)
- Animal feed (27.5%)
- Starch and starch based industry

Processed products like chips, sago and vermicelli made of tapioca are also popular in the country. Being easily digestible, it is an important ingredient in poultry and cattle- feeds. It is also widely used for production of industrial alcohol, starch and glucose. As cassava has no determined harvest time after which it spoils, farmers can have a staggered harvesting rather than a set date. This adds to the advantage in cassava based cropping system. Being sprout and grow as a new plant and thus enhance the multiplication ratio. It is based on the concept that once the bud sprouts, the roots developed would start drawing nutrients from the soil and no more from the mother planting material and therefore the size of planting material actually does not matter as far as sprouting is concerned.

Origin and Distribution: Cassava is not known in wild state. North-Eastern Brazil is the centre of origin. Cassava (*Manihot esculenta* Crantz) popularly known as tapioca, is a native of Brazil in Latin America and was introduced to India (Kerala) by the Portuguese in the 17th century. The evolutionary history of cassava is somewhat speculative due to the complex taxonomy and rare survival of fleshy plant parts, for subsequent archeological discovery. It is widely distributed in low land tropics of South and Central America. To Indian sub continent, the crop was introduced by Portuguese during 18th century. India made importation directly from South America in1794 and West Indies in 1840 (Cock, 1985). Portuguese distributed the crop from Brazil to countries like Indonesia, Singapore, Malaysia and India. Nigeria is the major growing country in world accounting for 50% of area and production. In India crop is cultivated in southern peninsular region, particularly Kerala, Tamil Nadu and Andhra Pradesh contributing 93% of area and 98% of production in the country. Kerala accounts for nearly 50% of total area under cassava in India and is mainly grown as rainfed crop.

Botany: Cassava belongs to family *Euphorbiaceae* and is diploid (2n=36). Polyploids with 2n=54 and 72 are also available. It is a perennial shrub producing 5 to 10 cylindrical tubers per plant. Being a member of family *Euphorbiaceae*, it produces latex. The stem is woody and variously branched. Two distinct types are present, one without branching at the top and the other with spreading nature. Leaves are palmately lobed with 5 to9 lobes.

Cassava is monoecious in nature and cross-pollinated. Female flowers are a few in numbers and are borne in the base of inflorescence and male flowers are borne above. Female flowers open about 10 days before male flower anthesis. Stigma is receptive from 6.30 a.m. and continues up to 2.30 p.m. Plants when raised from seeds produce typical tap root system. Since crop is mainly propagated by vegetative means by stem cuttings, numerous adventitious roots develop, of which a few develop into tubers. Tubers are composed of a thin peridium, white or purple cortex known as rind and central massive flesh rich in starch (25-40%). Bitterness often encountered in a few varieties and at certain stage is due to a bitter principle cyanogenic glucoside (HCN).

Germplasm Collection: At least 28 countries are known to maintaining germplasm.

Few important centers maintaining cassava germplasm are

1. CIAT, Colombia-5035 acc.
2. EMBRAPA, Brazil-1960 acc.
3. IITA, Nigeria-1286 acc. and Indonesia-700 acc.

Crop Improvement

High amount of cross-pollination results in heterozygous nature. In India crop improvement was taken up at the erstwhile Travancore University in 1940. Ever since the central tuber crop research institution was established at Trivandrum in 1963, significant achievements in varietal improvement of cassava were achieved. The main cassava breeding methods employed at CTCRI (ICAR-Central Tuber Crops Research Institute, Thirwanantapuram) were:

1. Selection from indigenous and exotic germplasm
2. Intervarietal hybridization

1. **Selection from the germplasm**: From a meager collection of 75 types in 1950, the germplasm has grown to 1465 consisting of 701 indigenous types, 696 exotics, 60 hybrids and 8 wild species at CTCRI. The evaluation of the introduced clones in the earlier years by botany department of the former Travancore University resulted in selection of M-4, which is popular even today due to its excellent culinary qualities. Recently another selection 'Sree Prakash' from germplasm (acc. S-856) was identified and released for its early maturing habit, coupled with good yield and acceptable qualities.
2. **Intervarietal hybridization:** As a result of large scale intervarietal hybridization of selected parents from the germplasm, CTCRI produced and released hybrid clones, H-97, H-165 and H-226 in 1971. Later two high yielding hybrids, Sree Visakham (H-1687) and Sree Sahya (H-2304) were released in 1974.
3. **Polyploidy breeding**: Though induced tetraploid did not show superior yield performance, the triploids developed by crossing tetraploid with diploids have shown superiority over tetraploids. A few of selected triploids have higher dry matter content and extractable starch content.
4. **Mutation breeding:** Radiation has been used to increase genetic variability in cassava. As a result of this, short-petiole mutant and chlorophyll mutant are being evaluated in CTCRI.

Achievements

Varieties differ in colour of rind and flesh, size of tubers, colour of stem, leaf and petiole, branching pattern, duration of crop and resistance to mosaic disease.

Varieties/Hybrids Released from CTCRI

M-4: It is a table purpose variety, introduced from Malaya, tubers having excellent culinary quality with low HCN content. It yields 20 to 25t/ha in 10 months. It is highly field tolerant to Cassava Mosaic Disease (CMD), but susceptible to mites.

Kalikalan: It is early and has good quality, well adapted to low fertility but highly susceptible to CMD.

Karuthakaliyan: It is spreading type and has non bitter tubers and yields 32 t/ha in 10 months. It is susceptible to CMD, bacterial leaf spot and mites.

H-97: It is hybrid between an indigenous cultivars chadayamangalam and exotic selection of Brazilian origin. The variety is characterized by light sepia colour of the merging leaf and conical medium sized tuber with 27 to 29% starch on fresh weight basis. It yields 25 to 35t/ha in 10 months. It is tolerant to CMD.

H-165: It is the hybrid between Chadayamangalam Vella and Kalikalan. Generally non branching with pinkish petiole and light brown emerging leaves. Tuber skin is golden brown and flesh is creamy with starch content of 23 to 25%. It yields 33 to 38t/ha in 8 to 9 months and tolerant to CMD. It is cultivated in large area of Tamil Nadu and Andhra Pradesh for industrial use.

H-226: It is a cross between M-4 introduced from Malaya and local cultivar Ethakkakaruppan. It has tubers with green colour rind and white flesh. Due to its very high starch content (28-30%) in tubers, this hybrid is very popular in industrial use in Tamil Nadu and Andhra Pradesh.

Sree Visakham (H-1687): It is a hybrid between an indigenous cultivar and an acc. from Madagascar. Tuber has brown rind and cream coloured flesh with 25 to 27% starch and carotene content is 466 IU/gm with low HCN. It yields 35 to 38t/ha and it is tolerant to CMD.

Sree Sahya (H-2304): This is hybrid resulted from crossing of five parents of which 2 are exotic and 3 are indigenous. Plants are tall, predominantly non branching, producing long necked tubers. Tuber skin is light brown and flesh is white with starch content of 29 to 31% and non bitter. It yields 35 to 40t/ha in 10-11 months and is tolerant to CMD.

Sree Prakash: This is a short, non-branching, early maturing (7-8 months) variety developed by clonal selection. It is highly tolerant to *Cercospora* leaf spot and yields 35 to 40 t/ha in 7 to 8 months. It is the selection from indigenous collection. Plants are relatively dwarf with good leaf retention and tubers are having good cooking quality and non bitter. It is suitable for single cropped rice field of Kerala.

Sree Harhsa (2-14): This is a triploid clone developed by crossing a diploid with an induced tetraploid clone of 'Sree Sahya'. Plants are stout, erect and non-branching with tubers of good cooking quality and high starch content (38-41%). It yields 35 to 40 t/ha in 7 to 8 months. It is derived from cross of triploid FOP-4 (2x) X H-230 (4x)}, released from CTCRI, Trivandrum in 1996. Mature stem is grayish and green with purple tinge colour of shoots. Leaves are thick broad with acuminate tip having light purple colour of emerging leaf. Tubers are conical in shape and light brown in colour. It is drought tolerant, robust plant and has high starch content acceptable especially for industrial use. It is susceptible to cassava mosaic disease and field tolerant to *Cercospora* leaf spot, spider mite and scale insect.

Sree Jaya (CI-649): It is selection from indigenous germplasm collection of cassava from Kottayam district, Kerala State and released from State Variety Release Committee, Kerala in 1998. It is early maturing (6-7 months) variety. Plants are erect and branching having reddish brown mature stem colour. Leaves are broad with light purple petiole. This variety is especially suitable for low land cultivation as a rotation crop in paddy based cropping system. The yield potential of this variety is 260-300 q/ha. However, this variety can also give up to 580q/ha under good cultural practices. This variety is moderately susceptible to cassava mosaic disease and field resistant to scale insect.

Sree Praba (TCH-2): This variety has been developed through pedigree method of breeding and released in 2000. Plants are medium in height and semi-spreading with light green of mature stem. Leaves are medium size, lanceolate and acuminate tip. Tubers are conical in shape with brown in colour. This variety is susceptible to cassava mosaic disease, field tolerant to *Cercospora* leaf spot and field resistant to scale insect. An average yield of this variety is 400 to 450 q/ha in 8 to 10 months of crop duration. This variety is suited for cultivation in Kerala, Tamil Nadu, Karnataka and Andhra Pradesh.

Sree Rekha (TCH-1): This variety has been developed through pedigree methods of breeding and released in 2000. Plants are erect branching having brownish white mature stem. Leaves are broad, lanceolate and acuminate on tip. Tubers are long conical with light brown skin colour. This variety is suitable for cultivation in upland and low land conditions and in non-traditional areas of Tamil Nadu, Karnataka and Andhra Pradesh. This variety is susceptible to cassava mosaic disease and field resistant to scale insect. Average tuber yield is 450 to 480 q/ha in 8 to 10 months.

Sree Vijaya (CI-731): It is a selection from indigenous collection of cassava from Thiruvanathapuram district and released for area of Kerala State during 1998. Plants are erect branching having greenish brown mature stem with green colour shoot. Leaves are broad with light green colour petiole. Tubers are conical in shape with brown skin colour. The average yield potential of this variety is 250 to 280 q/ha. This variety is field tolerant to *Cercospora* leaf spot and moderately susceptible to cassava mosaic disease. It is susceptible to spider mite and scale insect.

Three varieties viz., Nidhi, Kalpaka and Vellayani Hraswa were developed by Kerala Agricultural University.

Table: High yielding varieties released from CTCRI

S.No.	Variety	Average yield (t/ha^{-1})	Special Attributes
Cassava			
1	H-97	25-36	Drought Tolerance
2	H-165	33-38	Popular in industrial belt, duration: 8-9 month, Starch: 33- 38%.
3	H-226	30-35	Popular in industrial belt, Starch: 28-30%.
4	Sree Sahya	35-40	Hardy and highly resistant to drought
5	Sree Visakham	36-38	Rich in carotene
6	Sree Prakash	30-35	Early maturing (7-8 months) and shallow bulking.
7	Sree Harsha	35-40	Triploid high starch content (38-41%)
8	Sree Jaya	26-30	Early maturing (6-7 months)
9	Sree Vijaya	25-28	Early maturing (6-7 months)
10	Sree Rekha	45-48	Suited to both upland and low land cultivation.
11	Sree Prabha	40-45	Suited to both upland and low land cultivation.

Nidhi : It is high yielding (25.1 t/ha) and tolerant to mosaic disease. It is short duration (5 to 6 months) variety with grayish white stem and petiole white with red shade and skin is light pink.

Kalpaka (KMC-1): It is high yielding (28.4 t/ha) variety with short duration (6 months). It has non branching stem and pink tuber rind.

Vellayani Hraswa : It is high yielding (44.01 t/ha) variety with short duration (5 to 6 months). It has pink tuber rind and excellent cooking quality.

Tamil Nadu Agricultural University also developed three varieties viz., CO-1, CO-2 and CO-3.

CO-1 : It is a clonal selection with tubers having whitish brown skin, creamy rind and 35% starch. It yields 30 t/ha in 8 to 9 months.

CO-2: It is a branching variety with tubers having brown skin, creamy white rind and 34.6% starch. It yields 35 t/ha in 8 to 9 months.

CO-3: It is a branching variety having tubers with brown skin and 35.6% starch. It yields 42.6 t/ha in 8 months.

CO (Tp) 4: It yields 50 t/ha with starch content of 40 per cent. It is field tolerant to red spider mite and scales

CTCRI CO (Tp) 5: Released during 2007, yield potential is 30t/ha. Its starch content is 26 per cent and is resistant to CMD. It is suitable for irrigated conditions.

Selected References

Abdul, N. M., Joseph, J. K. and Karuppaiyan, R., 2004. Evaluation of okra germplasm for fruit yield, quality and field resistance to yellow vein mosaic virus. Indian J. Plant Genet. Resour., 17:241–244.

Alegbejo, M., Ogunlana, M. and Banwo, O., 2008. Short communication. Survey for incidence of Okra mosaic virus in northern Nigeria and evidence for its transmission by beetles. Span J. Agric. Res., 6:408- 411.

Ali, M., Hossain, M. Z., Sarker, N. C., 2000. Inheritance of yellow vein mosaic virus (YVMV) tolerance in a cultivar of okra (*Abelmoschus esculentus* (L.) Moench). Euphytica. 111:205–209.

Ali, S., Khan, M. A., Habib, A., Rasheed, S. and Iftikhar, Y., 2005. Correlation of environmental conditions with okra yellow vein mosaic virus and Bemisia tabaci population density. Int J. Agric. Biol., 7:142–144.

Allard R. W., 1960. Principles of Plant Breeding by Publishers: John, Wiley and Sons, Inc.

Anonymous, 1990. Report on international workshop on okra genetic resources; 1990 October 8–12; National Bureau for Plant Genetic Resources, New Delhi, India. Rome (Italy): IBPGR.

Aparna, J., Srivastava, K. and Singh, P. K., 2012. Screening of Okra Genotypes to Disease Reactions of Yellow Vein Mosaic Virus under Natural Conditions VEGETOS., 25 (1): 326-328.

Basset, M.J. (Ed.). 1986. *Breeding Vegetable Crops*. AVI Publ.

Bhattacharya, M.K., Nandpuri, K.S. and Singh, S., 1979. Genetic divergence in tomato. *Acta Horticulturae*, 93:289-300.

Bose, T.K., Kabir, J., Maity, T.K., Parthasarathy, V.A. & Som, M.G., 2003. *Vegetable Crops*. Vols. I-III. Naya Udyog.

Brewster, J.L., 1994. *Onions and other Vegetable Alliums*. CABI. FFTC. *Improved Vegetable Production in Asia*. Book Series No. 36.

Brunt, A., Crabtree, K., Gibbs, A., 1990. Viruses of tropical plants. Wallingford: CAB International.

Chauhan, D. V. S., 1972. Vegetable production in India. 3rd ed. Agra: Ram Prasad and Sons.

CTCRI, Tiruvananthapuram Publications in Website

Dhankhar, B. S. and Mishra, J. P., 2004. Objectives of okra breeding. In: Singh, P. K., Dasgupta, S. K., Tripathi, S. K., editors. Hybrid vegetable development. Binghamton, N. Y: Haworth Press. p. 195–209.

Dhankhar, B. S., Saharan, B. S., Sharma, N. K., 1999. 'Hisar Unnat': new YVMV resistant okra. Indian Hort., 44:2–6.

Dhankhar, B. S., Saharan B. S. and Pandita, M. L., 1997. Okra 'Varsha Uphar' is resistant to YVMV. Ind Hortic. 41:50–51.

Dhankhar, S. K., Dhankhar, B. S. and Yadava, R. K., 2005. Inheritance of resistance to yellow vein mosaic virus in an interspecific cross of okra (*Abelmoschus esculentus*). Indian J. Agric. Sci., 75:87–89.

Dhillon, B. S., Tyagi, R.K., Saxena, S. & Randhawa, G.J., 2005. *Plant Genetic Resources: Horticultural Crops*. Narosa Publ. House.
Dutta, O. P., 1984. Breeding okra for resistance to yellow vein mosaic virus and enation leaf curl virus. Annual Report Bangalore (India): IIHR.
Fageria, M.S., Arya, P.S. & Choudhary, A.K., 2000. *Vegetable Crops: Breeding and Seed Production*. Vol. I. Kalyani.
Falconer, D. S. and Trudy F.C. Mackay, 1996. Introduction to Quantitative Genetics. Fourth edition. Longman Publishers
Fauquet, C. M. and Stanley, J., 2005. Revising the way we conceive and name viruses below the species level: a review of geminivirus taxonomy calls for new standardized isolate descriptors. Arch. Virol., 150:2151–2179.
Fraser, R. S. S., 1990. Disease resistance mechanisms. In: Mandahar, C. L., editor. Plant viruses, II. Boca Raton, FL: CRC Press. p. 321–345.
Gardner, E.J., Simmons, M.J. and Snustad, DP., 1991. Principles of Genetics ,Publishers: John, Wiley and Sons, Inc.
Ghosh, S.P., Ramanujam, T., Jos, J.S., Moorthy, S.N. & Nair, R.G., 1988. *Tuber Crops*. Oxford & IBH.
Harihar Ram, 1998. Vegetable breeding – Principles and practices, Kalyani Publishers, Ludhiana.
Jambhale, N. D., Nerkar, Y. S., 1981. Inheritance of resistance to okra yellow vein mosaic disease in interspecific crosses of Abelmoschus. Theor. Appl. Genet., 60:313–316.
Jones, H.A. and L.K. Mann, *Onion and their allied* –World Book Series, Leonard Hill Books Ltd, London.
K.L. Chadha (Chief Editor), Advances in horticulture, Vol. 5 (vegetable), 7 (potato), 8 (tuber crops), 9 (spices), 11 (media), 12 (aromatic and ornamental plants)
Kallo, G. & Singh, K. (Ed.). 2001. *Emerging Scenario in Vegetable Research and Development*. Research Periodicals & Book Publ. House.
Kallo, G. (Ed.), Genetic improvement of tomato –Monograph on TAG 14.
Kalloo, G., 1998. *Vegetable Breeding*. Vols. I-III (Combined Ed.). Panima Edu. Book Agency.
Kulkarni, C.S., 1924. Mosaic and other related diseases of crops in the Bombay Presidency. Poona Agriculture College Magazine.
Kumar, J.C. and Dhaliwal, M.S., 1990. *Techniques of Developing Hybrids in Vegetable Crops*. Agro Botanical Publ.
Kurup, G.T., Palanisami, M.S., Potty, V.P., Padmaja, G., Kabeerathuma, S. & Pallai, S.V., 1996. *Tropical Tuber Crops, Problems, Prospects and Future Strategies*. Oxford & IBH.
Nariani, T. K. and Seth, M. L., 1958. Reaction of Abelmoschus and Hibiscus species to yellow vein mosaic virus. Indian Phytopathol., 11:137–140.
Nerkar, Y. S. and Jambhale, N. D., 1985. Transfer of resistance to yellow vein mosaic from related species into okra (*Abelmoschus esculentus* (L.) Moench). Indian J. Genet., 45:261–270.
Paroda, R.S. & Kalloo, G. (Eds.). 1995. *Vegetable Research with Special Reference to Hybrid Technology in Asia-Pacific Region*. FAO.
Peter, K.V. and Pradeep Kumar, T. 2008. *Genetics and Breeding of Vegetables*. (Revised Ed.). ICAR.
Phadvibulya, V., Puripanyavanich, V., Adthalungrong, A., Kittipakorn, K. and Lavapaurya, T., 2004. Induced mutation breeding for resistance to yellow vein mosaic virus in okra., In: Proceedings of a Final Research Coordination Meeting organized by the Joint FAO/IAEA Division of Nuclear Techniques in Food and Agriculture; 2003 May 19–23; Pretoria (South Africa) p. 155–175.
Prabu, T. and Warade, S. D., 2009. Biochemical basis of resistance to yellow vein mosaic virus in okra. Veg. Sci. 36(3 Suppl.): 283-287.

Prohens, Jaime and Nuez Fernando, 2008. Vegetables I & II, Springer International Edition.

Pullaiah, N., Reddy, T. B., Moses, G. J., Reddy, B. M. and Reddy, D. R., 1998. Inheritance of resistance to yellow vein mosaic virus in okra (*Abelmoschus esculentus* (L.) Moench). Indian J. Genet. Plant Breed., 58:349–352.

Pun, K.B. and Doraiswamy, S., 1999. Effect of age of okra plants on susceptibility to okra yellow vein mosaic virus. Indian J. Virol., 15:57–58.

Rai, N. & Rai, M., 2006. *Heterosis Breeding in Vegetable Crops*. New India Publ. Agency.

Ram, H.H., 2001. *Vegetable Breeding*. Kalyani.

Rashid, M. H., Yasmin, L., Kibria, M. G., Mollik, A. K. M. S. R. and Monowar-Hossain, S. M., 2002. Screening of okra germplasm for resistance to yellow vein mosaic virus under field conditions. Plant Pathol. J., 1:61–62.

Rick, C.M., 1969. Origin of cultivated tomato, current status and the problem. *Abstract*, XI. International Botanical Congress, p. 180.

Rick., C.M., 1978. The tomato. *Scientific American*, 239(2): 76-87.

Salehuzzaman, M., 1985. Screening of world germplasm (*Abelmoschus esculentus* L.) for resistance to yellow vein mosaic virus. Bangladesh J. Agric., 10:1–8.

Salehuzzaman, M., 1986a. Breeding of yellow vein mosaic virus resistant okra. In: Proceedings of the 11th Annual Bangladesh Science Conference, Bangladesh Association for the Advancement of Science; 1986 Sep 20–24; Dhaka (Bangladesh). p. 8–9.

Salehuzzaman, M., 1986b. Nature of resistance to yellow vein mosaic virus in okra [of West Africa and Bangladesh]. In: Proceedings of the 11th Annual Bangladesh Science Conference, Bangladesh Association for the Advancement of Science; 1986 Sep 20–24; Dhaka (Bangladesh). p. 9–10.

Sastry, K. S. M. and Singh, S. J., 1974. Effect of yellow vein mosaic virus infection on growth and yield of okra crop. Indian Phytopathol., 27:294–297.

Sharma, B. R. and Dhillon, T. S., 1983. Genetics of resistance to yellow vein mosaic virus in interspecific crosses of okra. Genet. Agraria., 37:267–275.

Sharma, B. R. and Sharma, D. P., 1984a. Breeding for resistance to yellow vein mosaic virus in okra. Indi J Agric Sci. 54:917–920.

Sharma, B. R. and Sharma, O. P., 1984b. Field evaluation of okra germplasm against yellow vein mosaic virus. Punjab Hort. J., 24:131–133.

Sharma, B. R., Kumar, V. and Bayay, K. L., 1981. Biochemical basis of resistance to yellow vein mosaic virus in okra. Genet. Agraria., 35:121–130.

Simmonds, N.W., 1976. Evolution of crop plants, Longman, London

Singh, P.K., Dasgupta, S.K. and Tripathi, S.K., 2004. *Hybrid Vegetable Development*. International Book Distributing Co. Book Distributing Co.

Singh, B., Rai, M., Kalloo, G., Satpathy, S. and Pandey, K. K., 2007. Wild taxa of okra (*Abelmoschus spp.*): reservoir of genes for resistance to biotic stresses. Acta Hort., 752:323–328.

Singh, H. B., Joshi, B. S., Khanna, P. P. and Gupta, P. S., 1962. Breeding for field resistance to yellow vein mosaic in bhindi. Indian J. Genet., 22:137–144.

Solakey, S. S., Akhtar, S., Kumar, R., Verama, R. B. and Sahajan, K., 2014. seasonal response of okra (*Abelmoschus esculentus* L. Moench) genotypes for okra yellow vein mosaic virus incidence. African J. Biotech., 13 (12): 1336-1342.

Swanson, M. M. and Harrison, B. D., 1993. Serological relationships and epitope profiles of isolates of okra leaf curl geminivirus from Africa and the Middle East. Biochimie., 75:707–711.

Thakur, M. R., 1976. Inheritance of resistance to Yellow Vein Mosaic (YVM) in a cross of okra species, *Abelmoschus esculentus, A. manihot ssp. manihot*. SABRAO J., 8: 69–73.

Usha, R., 2008. Bhendi yellow vein mosaic virus. In: Rao, G. P., Kumar, P. L., Holguin-Pena, R J., editors. Characterization, diagnosis and management of plant viruses. Vol. 3. Houston, TX: Studium Press. p. 387–392.

Vashisht, V. K., Sharma, B. R. and Dhillon, G.S., 2001. Genetics of resistance to yellow vein mosaic virus in okra. Crop Improv., 28:18–25.

Vavilov, N.I., 1926. Studies on the origin of cultivated plants. *Bulletin of Applied Botany*, 16: 2.